Astrophysics and Space Science Proceedings

For further volumes:
http://www.springer.com/series/7395

The Square Kilometre Array: Paving the way for the new 21st century radio astronomy paradigm

Proceedings of Symposium 7 of JENAM 2010

Editors

Domingos Barbosa
Sonia Anton
Leonid Gurvits
Dalmiro Maia

Springer

Sociedade Portuguesa de Astronomia

eas

Editors
Domingos Barbosa
Radio Astronomy Group,
Instituto de Telecomunicações, Campus
Univ. de Aveiro
Aveiro
Portugal
dbarbosa@av.it.pt

Sonia Anton
CICGE
Departamento de Matemática
Rua do Campo Alegre 687
4169-007 Porto
Portugal
santon@fc.ul.pt

Leonid Gurvits
The Joint Institute for VLBI in Europe
Dwingeloo
Netherlands
lgurvits@jive.nl

Dr. Dalmiro Maia
CICGE-UP
Observatorio Astronomico
Manuel de Barro
Alameda do Monte da Virgem
4430-146 Vila Nova de Gaia
Portugal
dmaia@fc.up.pt

ISSN 1570-6591
ISSN 1570-6605 (eBook)
ISBN 978-3-642-44532-3
ISBN 978-3-642-22795-0 (eBook)
DOI 10.1007/978-3-642-22795-0
Springer Heidelberg Dordrecht London New York

Printed on acid-free paper

Springer is part of Springer Science+Business Media (www.springer.com)

Preface

The Square Kilometre Array will provide more than one order of magnitude improvement in sensitivity compared with any existing radio telescope over a wavelength range of several hundred to one, from decametric to microwave wavelengths. It will revolutionize the study of the most abundant element in the Universe, hydrogen, from the epoch of reionization to the present day, probing the onset formation period of the very first stars, look in depth to proto-planets and, through the precision timing of pulsars, detect the distortions of space-time due to gravitational radiation. SKA is a sensor machine spawning 3,000 km in extension and a collecting area of more than 1 km^2, using technologies of twenty-first century. SKA will allow the study at radio wavelengths of a wide range of phenomena initially studied at other wavelengths as well as opening a new discovery window on new phenomena at radio wavelengths, as has been the case with radioastronomy whenever a jump in sensitivity happened. The JENAM SKA Symposium was aimed at bringing these diverse opportunities to the attention of both theoretical and observational astronomers working at all wavelengths, including the potential for synergies with other facilities. But SKA represents also a technological challenge with recognized societal impacts on Information and Computing Technologies and Green Energy, as it was demonstrated at the COST conference in Rome, 2010. This Symposium was timely as Portugal is well suited to provide key insight on these aspects of new detector technologies and renewable energy for this project by hosting the European Site Emulator for the Aperture Array Verification Program, providing a testing platform for some of the developments for the Advanced Instrumentation program of the SKA.

We therefore brought to the attention of the broader astronomical community the scientific potential of the SKA by discussing its scientific priorities and their impact on the design of the whole project, in a crucial year for the project. We explored also the synergies between the SKA and other next generation astronomical facilities in different wavelength domains such as the ALMA, ELTs, LSST, JWST, GRE, IXO,

Gaia and Euclid and high energy facilities (Auger) in the true spirit of JENAM events. To this endeavor we warmly thank Ken Kellerman (NRAO) for helping us shape the program and all the support from the European SKA Consortium (ESKAC).

Aveiro	*Domingos Barbosa*
Porto	*Sonia Anton*
Dwingeloo	*Leonid Gurvits*
Porto	*Dalmiro Maia*

SYMPOSIUM 7: The Square Kilometre Array: Paving the way for the new 21st century radio astronomy paradigm

Scientific Organising Committee

Prof. K. I. Kellermann, NRAO, USA (chair)
Prof. R. Bachiller, IGN, Spain
Prof. D. de Boer, CSIRO, Australia
Prof. J. Cordes, Cornell, USA
Prof. P. J. Diamond, U. of Manchester, UK
Dr. L. Feretti, INAF - IRA, Bologna, Italy
Prof. M. A. Garrett, ASTRON, The Netherlands
Prof. L. I. Gurvits, JIVE, The Netherlands
Prof. J. M. van der Hulst, U. of Groningen, The Netherlands
Prof. J. Jonas, South Africa
Prof. S. R. Rawlings, U. of Oxford, UK
Prof. R. T. Schilizzi, SPDO
Dr. S. Torchinsky, Observatoire de Paris, France
Prof. Russ Taylor, U. Calgary, Canada
Prof. J. A. Zensus, MPIfR, Germany
Prof. A. van Ardenne, ASTRON, The Netherlands

Local Organising Committee

Dr. Domingos Barbosa, IT- Aveiro, Portugal, (chair)
Dr. Sónia Anton, CICGE-UP, Portugal
Dr. Luis Cupido, IPFN, Portugal
Dr. Mercedes Filho, CAUP, Portugal
Dr. Dalmiro Maia, CICGE-UP, Portugal
Dr. António Magalhães, CICGE-UP, Portugal
Dr. Mário Santos, CENTRA-IST, Portugal

Contents

The SKA Challenge

David R. DeBoer

Abstract The SKA is an international project to build a large radio telescope that has the potential to transform our understanding of the Universe over decades to come. As a massively parallel system in every aspect (technical, political, administrative, ...) it represents a major challenge to realize, despite the existence of the underlying technology. It also represents the opportunity to greatly advance our understanding of the Universe and drive technical innovation in a number of sectors.

1 Introduction

The Square Kilometre Array (SKA) will be one of the Great Observatories to answer fundamental questions about the Universe: its life history, its inner workings and our place in it. The SKA is an ambitious program in part because from the start it has been international in scope both as a principle of collaboration and to address its ambitious breadth. A central premise of the SKA is that market technology drivers have driven the performance and cost of the central technologies such that sufficient quantities can be affordably deployed to achieve the huge increase in sensitivity called for by the science. There have been many papers and memos written on the SKA and its potential science and technical prowess, which may be found in [2, 4, 11], the SKA Memo Series (www.skatelescope.org) and references therein.

Cost and complexity are the key SKA challenges. Performance of the core technology is certainly adequate in order to meet the design goals; however can it be done at an appropriate price per performance? Computing roadmaps predict that

D.R. DeBoer (✉)
Commonwealth Scientific and Industrial Research Organization (CSIRO), P.O. Box 76, Epping, NSW 1710, Australia
e-mail: ddeboer@berkeley.edu

D. Barbosa et al. (eds.), *The Square Kilometre Array: Paving the way for the new 21st century radio astronomy paradigm*, Astrophysics and Space Science Proceedings,
DOI 10.1007/978-3-642-22795-0_1,

appropriate high-performance computing will exist; however can it be afforded? The cost of operating the facility is a key constraint; if we can afford to build it can we afford to operate it?

The underlying cause of these issues is that the SKA is a massively parallel system in every arena: technically, operationally, organizationally, and politically. Although the goal is to keep the many parallel sub-systems simple (and inexpensive), the huge numbers on different sensor platforms in a complex international stakeholder environment introduces great complexity. Addressing these issues of costing and complexity head-on is needed to successfully deploy what could become one of the largest and most iconic science projects ever.

Two areas have been shortlisted by the international community for hosting the SKA:

Australia: The core site in the Murchison region of outback Western Australia, with sites across the Australian continent into New Zealand.

Southern Africa: The core site in the Karoo region of Northern Cape Province, with sites into Namibia, Botswana, Mozambique, Mauritius Madagascar, Kenya and Zambia.

This paper will provide a brief overview of the SKA, highlighting the challenges and opportunities.

2 The Square Kilometre Array

The science and technology of radio astronomy is a mature field and addresses many fundamental aspects of astrophysics and cosmology. Existing large facilities are highly productive, heavily used and in most cases decades old. Looking at future likely science priorities in most cases requires adding much more sensitivity and capability to those already there. A collecting area on the order of 1 km^2 has been proposed and is being pursued by an international collaboration for the Square Kilometre Array (SKA) [4, 11]. A rendering is shown in Fig. 1.

Fortunately the technologies needed lie within the current technological explosion usually generically lumped within the rubric of "Moore's Law". Also fortunately, radio astronomy techniques readily allow for implementations of many smaller identical units arrayed together – something in which industry excels. The numbers discussed for the SKA – 1000's to 1,000,000's of units, depending on the component – is relatively small for large industry, but is very large in the experience of radio astronomy. Some of the key specifications, largely taken from [10], are shown in Table 1.

The ability to array many smaller units together in a technique called radio interferometry allows great flexibility in constructing and commissioning of the telescope as well. That is, a smaller number of units may be built, tested and operated while the rest of the array is being constructed. This allows the commissioning process a head start and also allows significant science to flow early. For good or ill, this ability also provides an additional contingency factor. There is a strong

Fig. 1 A rendering of the core of the dish (array. Credit Swinburne Astronomy productions)

Table 1 Key SKA specifications

Frequency range	70–300 MHz	Low
Survey speed	300–3,000 MHz	Mid
	3–10 GHz	High
	10^7 to 10^{10} $m^4 K^{-2} deg^2$	Varies across band – these are bracketing values
Configuration	20% within 1 km 50% within 5 km 75% within 180 km 1,000% within 5,000 km	Multiple cores and roughly spiral arms

cautionary tale here though – to really see a savings in reducing the scope of the array it must be done fairly early in the process. Deciding late in the game to build out to 90% of the original number does not save 10% of the overall cost.

Another very important aspect is of course the cost of operating the facility – whole-of-life costing is central in understanding the SKA optimization and constraints. The cost of energy, which will likely exceed 100 MW for the entire scope of the SKA, may well be a limiting factor. Early thought however in energy supply-side, demand-side and operational-modes could enable its success, as well as drive innovation in that sector.

3 SKA Key Science

The SKA is a revolutionary step in understanding some of the most basics tenets of science. A good discussion of the science to be done is contained within the book [2]. A brief sketch of some of the highlights is given below.

3.1 Probing the Dark Ages

About 400,000 years after the Big Bang the hot universe underwent a phase change where the plasma soup recombined into primarily neutral hydrogen becoming mostly transparent to its radiation field. As this material coalesced into the first galaxies and stars they then started to re-ionize the gaseous material in the so-called "epoch of re-ionization" (or EoR) which finished when the Universe was about 1 billion years old. During this period, there were bubbles of ionized material in a sea of neutral material which actually allows for imaging of a 3-dimensional volume via mapping hydrogen as opposed to a final scattering surface as for the cosmic microwave background (CMB).

The goal for the SKA is to image this 3-D volume with red-shifted HI at 100–300 MHz, with an expectation that arrays currently being implemented will make the first detections and possibly the first crude maps. This is analogous to the development of the CMB, where the first detection was made by [8], the Cosmic Background Explorer (COBE) satellite making the first maps in 1990 [6], followed up by maps at higher resolution and precision from the Willkinson Microwave Anistropy Probe (WMAP) in 2003 [12]. The full SKA promises to shine a great light into the Dark Ages.

3.2 The Lives of Galaxies and Dark Energy

Most of the 4.6% of the baryonic mass of the Universe is hydrogen and understanding the Universe requires understanding the distribution of hydrogen throughout time, which is the domain of radio astronomy. The SKA will produce incredibly deep and detailed maps of hydrogen out to a redshift of about 2. This will also trace Dark Matter and produce a complete census of galaxies over redshift to constrain the equation of state of Dark Energy.

3.3 The Magnetic Life of the Universe

Threading the Universe at all spatial scales, magnetic fields have undoubtedly played a huge role in the evolution of stars and galaxies. These fields were generated after the Big Bang at some point in the early Universe via a mechanism yet to be cornered and understood. The SKA will measure Faraday rotation of deep background objects and will track the structure and evolution of magnetic fields over a large volume of the Universe's history.

3.4 Gravity and Relativity

Pulsars, and particularly millisecond pulsars, are nearly perfect mechanisms to probe the extreme physics which they inhabit: densities of up to $10^{14}\,\mathrm{g/cm^3}$, magnetic fields up to 10^{14} Gauss, and surface velocities approaching a good fraction of c. As the most precise clocks known, and sometimes with interesting orbital companions, they serve as celestial timekeepers to test theories of strong gravity and by precisely timing millisecond pulsars over a range of locations to potentially directly measure gravitational waves.

3.5 *The Cradle of Life*

At the frequencies at which the SKA will operate and with the sensitivity and scale of the final instrument, the SKA will be able to peer into dusty star-formation regions and potentially measure the direct impact of planetary formation in the system. The SKA will also be able to conduct sensitive searches for long-chain pre-biotic molecules. The sensitivity of the SKA will enable it to conduct searches for transmissions from other technical societies – for the first time at a level where it may be possible to detect incidental or "leakage" radiation, as opposed to deliberately targeted beacons.

And of course, with the dramatic increase in nearly all parameters, the SKA will advance mankind's legacy of exploring the unknown – its greatest discoveries are likely not things currently known.

3.6 *SKA Key Technology*

The key technology for the SKA is undergirded by the technological explosion that is the modern world, which increasingly drops the cost/performance of critical components. This is a central factor in the realization of the SKA. Moving most of the cost to "silicon not steel" is a key advantage in radio interferometry, which both lowers the cost and increases the performance. In addition, there have been huge gains in mid-scale production techniques which facilitates the production of the structural components helping to offset the rise in the cost of the raw materials. The technology can be thought to fall under the "three S's":

1. *Steel*: The raw materials to build the sensors and the infrastructure.
2. *Silicon*: The medium allowing the great advances in processing and computing that enable the realization, including in the analog electronics.
3. *Sun*: The energy needed to build and power the array – this is one of the key constraints for the success of the SKA.

These will be briefly discussed in a slightly different format below.

3.7 *Sensors*

Given the large frequency coverage for the SKA (roughly 200:1), multiple sensor platforms will be used. All are advantaged by using industrial-scale production techniques to cost effectively deploy collecting area and receptors.

Sparse Aperture Arrays: At metre-wavelengths (<300 MHz) the most cost-effective way to deploy collecting area is via many dipoles on the ground. This also allows great flexibility since the field-of-view accessible is very nearly the entire sky. The frequencies supported by these sensors include those for Epoch of Re-ionization science. This approach is being pursued by several groups around the world: in Europe principally via the International LOFAR Telescope (ILT: [9]) and in the US via the Hydrogen Epoch of Re-ionization Arrays (HERA: [1]).

Wide-Field Arrays: At decimetre-wavelengths (0.3–3.0 GHz) the need for very fast surveys for red-shifted hydrogen has led to the investigation of techniques for field-of-view-expansion. One approach is to place phased array feeds in the focal planes of parabolic dishes and increase the survey speed by a factor of the number of formed beams (of order 30). This is under investigation in Australia for the Australian SKA Pathfinder (ASKAP: [3]), in the Netherlands for the APERture Tile-In-Focus (APERTIF: [7]) project for the Westerbork Synthesis Radio Telescope, and in the US for arrays for the Greenbank Telescope (FLAG: [13]) and Arecibo.

Another approach is to dispense with the dishes and place tiles of receptors directly on the ground analogous to the sparse aperture arrays. These are arrayed in tight groups to fully sample the wavefront over many wavelengths and are hence called Dense Aperture Arrays. This increases the survey speed and flexibility, but requires the deployment of very many receiver chains. Constrained architectures are under investigation to produce a cost-effective yet still very flexible system. Work on this is led out of Europe previously under the SKA Design Studies program and currently under the Aperture Array Verification Program.

Dish Arrays: At centimetre-wavelengths (>3 GHz) single-pixel feeds on parabolic reflectors remain the most cost-effective to deploy sensitive collecting area to higher frequencies (up to about 20 GHz). These feeds could likely co-exist on the same antenna structures as the phased array feeds. Small, cooled phased array feeds may become advantageous to deploy instead at some time in the near future however multiple phased arrays feeds would need to be deployed since the bandwidths are currently limited to 2.5:1. Ultra-wideband single-pixel feeds and cost effective antenna structures are under development in the USA under the SKA Technology Development Program. South Africa under MeerKAT [5] and Australia under ASKAP are also developing deployed antenna systems.

3.8 Signal Processing and Transport

The data volume for the SKA is staggering and the ability to transport and handle this volume in "real-time" is a key component of successfully realizing the SKA. In the core every sensor of thousands could generate multiple Terabits-per-second (Tb/s) while the outlier stations could likely be limited to a more "moderate" 100 Gb/s but over thousands of kilometres.

Current processing systems tend to utilize Field-Programmable Gate Array (FPGA) technology as a cost- and energy-efficient means to flexibly handle the large data volumes. At scales of the SKA however, energy requirements will likely drive to Application Specific Integrated Circuit (ASIC) technology. This raw data will then be compressed in time (integrated) and/or frequency (averaged) to a rate appropriate for the post-processing computing.

3.9 Computing, Archiving and Curation

Once consolidated and correlated at the central site, the volume of raw data is too large to be saved and processed off-line, so real-time intelligent pipelined techniques are needed. High-Performance Computing (HPC) at peta-operations per second (Peta-ops) and archives with petabytes of memory will be needed. In addition, the ability to maintain and serve this data (curate) is required to actually produce the science outcomes. Projections of HPC show that the scale of the SKA requires one of the fastest computers at that future time, an ambitious specification.

3.10 Infrastructure and Energy

With the likely exception of the HPC, the SKA will be deployed in remote areas. Power and connectivity will be a major challenge, and will remain possibly the biggest constraint in the operational model of the SKA. Regarding power, the need is for $\sim$100 MW at the core at all times in the remote desert and $\sim$1 MW at several dozen sites across a continent. This argues for a large plant near the core to supply it and possibly the nearby stations, plus small "islanded" power stations, preferably stand-alone solar, for the antenna stations. Some may be near enough to grid power to be supplied that way.

Demand-side reduction is an absolutely key design criterion. Digital design must be low-energy architectures and novel HVAC schemes using geo-regulation or geothermal will be important.

The HPC will likely be near a large population centre, but will also be one of the largest power hurdles, so "green" metrics as well as speed metrics will be very important. As a huge issue for industry as well all these issues are being investigated at very large scales.

4 Conclusion

The Square Kilometre Array presents many challenges and opportunities scientifically and technically. The key challenges relate to the massively parallel nature of the SKA and the cost and complexity which may ensue. These challenges are not at all unique to radio astronomy and in fact aren't uniquely large relative to other sectors. As we know, industry deals with much larger scales.

For radio astronomy, some issues relate to less experience with this production and operational scale, which can be gleaned from industry. However the typical relative performance for a radio telescope does tend to differ and a close collaboration will be needed.

Surmounting these challenges provides a great potential to deliver innovation in analog, digital, HPC, energy, and signal transport systems domains. An SKA geared towards answering these big questions provides inspiration to all generations and promises to be an iconic instrument for our time.

References

1. Backer, D.C., Aguirre, J., Bowman, J.D., Bradley, R., Carilli, C.L., Furlanetto, S.R., Greenhill, L.J., Hewitt, J.N., Lonsdale, C., Ord, S.M., Parsons, A., Whitney, A.: "HERA: Hydrogen epoch of reionization arrays." White paper submitted to US ASTRO2010 Decadal Review. Reionization.org (2009)
2. Carilli, C., Rawlings, S. (eds): "Science with the Square Kilometre Array." Elsevier Press, UK (2004)
3. DeBoer, D.R., Gough, R.G., Bunton, J.D., Cornwell, T.J., Beresford, R.J., Johnston, S., Feain, I.J., Schinckel, A.E., Jackson, C.A., Kesteven, M.J., Chippendale, A., Hampson, G.A., O"Sullivan, J.D., Hay, S.G., Jacka, C.E., Sweetnam, A.W., Storey, M.C., Ball, L., Boyle, B.J.: "Australian SKA pathfinder: A high-dynamic range wide-field of view survey telescope." Proc IEEE. **97**(8):, 1507–1521 (2009)
4. Dewdney, P.E., Hall, P.J., Schilizzi, R.T., Lazio, T.J.W.: "The Square Kilometre Array". Proc IEEE **97**(8), 1482–1496 (2009)
5. Jonas, J.L.: "MeerKAT: The South African Array with composite dishes and wide-band single-pixel feeds." Proc IEEE. **97**(8), 1522–1530 (2009)
6. Mather, J.C., Cheng, E.S., Eplee, R.E., Jr., Isaacman, R.B., Meyer, S.S., Shafer, R.A.: "A preliminary measurement of the cosmic microwave background spectrum by the Cosmic Background Explorer (COBE) satellite." Astrophys. J. 2 Lett. **354**, L37–L40 (1990)
7. Oosterloo, T., Verheijen, M., van Cappellan, W., Bakker, L., Heald, G., Ivashina, M.: "Apertif: The Focal Plane Array System for the WSRT" in Wide Field Science and Technology for the Square Kilometre Array: Proceedings of the Final SKADS Conference (2009)
8. Penzias, A.A., Wilson, R.W.: "A measurement of excess antenna temperature At 4080 Mc/s". Astrophys. J. Lett. **142**, 419–442 (1965)
9. Rottgering, H.J.A., Braun, R., Barthel, P.D., van Haarlem, M.P., Miley, G.K., Morganti, R., Snellen, I., Falcke, H., de Bruyn, A.G., Stappers, R.B., Boland, W.H.W.M., Butcher, H.R., de Geus, E.J., Koopmans, L., Fender, R., Kuijpers, J., Schilizzi, R.T., Vogt, C., Wijers, R.A.M.J., Wise, M., Brouw, W.N., Hamaker, J.P., Noordam, J.E., Oosterloo, T., Bahren, L., Brentjens, M.A., Wijnholds, S.J., Bregman, J.D., van Cappellen, W.A., Gunst, A.W., Kant, G.W., Reitsma, J., van der Schaaf, K., de Vos, C.M.: "LOFAR - Opening up a new window on the Universe" in Cosmology, Galaxy Formation and Astroparticle Physics on the Pathway to the SKA: Proceedings of the Conference held in Oxford 10–12 April 2006 (2007)
10. Schilizzi, R.T., Alexander, P., Cordes, J.M., Dewdney, P.E., Ekers, R.D., Faulkner, A.J., Gaensler, B.M., Hall, P.J., Jonas, J.L., Kellerman, K.I.: Preliminary Specifications for the Square Kilometre Array, SKA Memo 100 (2007)
11. Schilizzi, R.T., Dewdney, P.E., Lazio, T.J.W.: "The Square Kilometre Array." Proc. SPIE **7733**, 18 (2010)
12. Spergel et al: Astrophys. J. Supplement **148**, 175–194 (2003)
13. Warnick, K.F., Carter, D., Webb, T., Landon, J., Elmer, M., Jeffs, B.D.:"Design and characterization of an active impedance matched low noise phased array feed." IEEE Trans. Antennas Propagat. in review (2009)

The SKA New Instrumentation: Aperture Arrays

A. van Ardenne, A.J. Faulkner, and J.G. bij de Vaate

Abstract The radio frequency window of the Square Kilometre Array is planned to cover the wavelength regime from cm up to a few meters. For this range to be optimally covered, different antenna concepts are considered enabling many science cases. At the lowest frequency range, up to a few GHz, it is expected that multi-beam techniques will be used, increasing the effective field-of-view to a level that allows very efficient detailed and sensitive exploration of the complete sky. Although sparse narrow band phased arrays are as old as radio astronomy, multi-octave sparse and dense arrays now being considered for the SKA, requiring new low noise design, signal processing and calibration techniques. These new array techniques have already been successfully introduced as phased array feeds upgrading existing reflecting telescopes and for new telescopes to enhance the aperture efficiency as well as greatly increasing their field-of-view (van Ardenne et al., Proc IEEE 97(8):2009) by [1]. Aperture arrays use phased arrays without any additional reflectors; the phased array elements are small enough to see most of the sky intrinsically offering a large field of view.

The implementation requirements of high frequency, astronomically capable phased arrays are severe in terms of power and cost due to the large numbers of channels and the amount of digital processing required. However, technological roadmapping shows that a cost effective large scale implementation for the SKA is achievable soon. An aperture array covering this frequency range is the only instrument able to perform some of the most challenging science experiments planned for the SKA and is likely to make some transformational discoveries. In the context of defining and developing the next SKA phase the international Aperture Array Verification Program, is working on both the sparse low frequency array from 70 to 450 MHz and a dense array from 400 to 1,450 MHz as the low frequency system for the SKA.

A. van Ardenne (✉)
ASTRON, P.O. Box 2, 7990 AA Dwingeloo, The Netherlands
e-mail: ardenne@astron.nl

D. Barbosa et al. (eds.), *The Square Kilometre Array: Paving the way for the new 21st century radio astronomy paradigm*, Astrophysics and Space Science Proceedings,
DOI 10.1007/978-3-642-22795-0_2,

The work aims to provide insight into the status of enabling technologies and technical research on polarization, calibration and side lobe control required to fully realise the potential of phased arrays for the SKA aperture synthesis array.

1 Introduction

The Square Kilometre Array, SKA is the next generation low frequency radio telescope with preliminary specifications described by Dewdney et al. [2]. The work performed in the European funded FP6 programme, SKA Design Studies, SKADS [3, 4], showed that an implementation of the SKA using phased aperture arrays, AAs, operating from 70 MHz up to 1.4 GHz with a dish based array covering ~1.2–10 GHz represents the most capable design for the agreed SKA Phase 2 science case [5]. Here we discuss the scientific benefits of AA and the need for the proposed frequency range; the development and trade-offs of a suitable array; and the system implications including central processing load.

The deployment of the SKA, starts with an ~10% instrument, Phase 1, in 2016 includes a low frequency sparse AA covering 70–450 MHz, the details are described in SKA memo 125 [6]. The development of the technically and astronomically less mature dense AA system will continue in parallel preparing for deployment in SKA Phase 2 commencing in 2018. This schedule enables SKA Phase 1 to benefit from the experience gained with current low frequency AAs, LOFAR [7] and MWA [8] systems, followed by the more challenging dense AA in Phase 2. These developments are core to the Aperture Array Verification Program [4, 9] advancing AA's for the SKA.

2 Scientific Benefits

An AA is a very different receiver concept from conventional dishes; it has multi-beaming capabilities, with very fast changes in observation direction. If we consider the benefits to the SKA Design Reference Mission [5], Figure shows the sensitivity and survey speed requirements of the various science experiments. Most of the major surveys are performed below the neutral hydrogen line at 1.4 GHz; this sets the highest frequency required of the AA system.

Remote galaxies are receding progressively faster due to the expansion of the Universe, this Doppler shifts the hydrogen line to lower frequencies, which enables measuring three dimensional structure, however, the signal also gets progressively weaker and increasing the survey speed requirement at lower frequencies, it was shown e.g. in Alexander [5] that the FoV needs to increase as λ^3 for an optimum survey time. An AA can adjust survey speed as a function of frequency by varying the number of beams across the bandwidth.

The requirement for high dynamic range, particularly for continuum experiments, exceeding $10^7:1$ will be very difficult to achieve. AA's have the characteristics necessary to achieve excellent dynamic range: physical stability, unblocked aperture, small individual beams due to large diameter of the array, ability for exquisite calibration of the "surface" over frequency. There is substantial research ongoing into optimizing the appropriate calibration techniques.

Pulsar and transient detections require a large number of beams for fast surveys and timing of multiple pulsars concurrently; AAs also have the ability to buffer data which enables a "look back" to find the pre-cursor of a transient event.

A subset of Fig. 1 could similarly be produced for the first SKA phase using the detailed science Design Reference Mission has been produced [10].

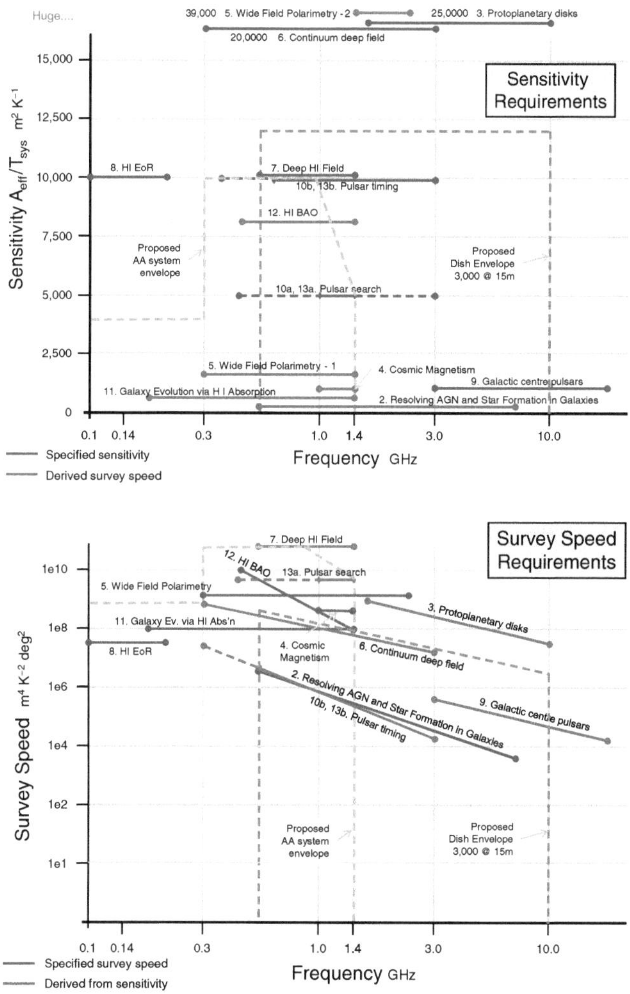

Fig. 1 The sensitivity and survey speed requirements for the SKA science case. The sensitivity specifications are shown in the left hand panel with the survey speed's on the right. Note that high survey speeds are typically needed below 1.4 GHz

3 Designing for Aperture Arrays

The production of SKA_1 will start in 2016. In order to have a production-ready design of SKA-low, a series of Aperture Array Verification Systems (AAVS) will be built, starting with a relatively open exploration phase and subsequently focusing on the final design. Three phases have been identified, starting with AAVS0, a modest antenna test system of approximately ten antennas, AAVS1 with 250 m^2 collecting area, similar in size to the first LOFAR initial test stations or the MWA 32 Tile system, and AAVS2. AAVS1 will demonstrate electromagnetic and front-end performance with sufficient collecting area in order to make astronomical verification possible, and should establish the antenna tile and station configuration. AAVS1 will be commissioned by the end of 2012 or early 2013. AAVS2 is the SKA_1 pre-production array, built with production tooling and sufficiently large area (1–2% SKA_1). The AAVS1 and AAVS2 will be build at the chosen SKA site or at a site, similar in terms of RFI and climate conditions i.e. in Portugal near Moura.

For AA-low, several antenna element types are being evaluated including options of splitting the frequency band in two This might be useful to limit the required bandwidth of each antenna but, importantly a single antenna array with e.g. an $\lambda/2$ frequency of 130 MHz will be very sparse at the top-end of the band, resulting in a low filling factor and many grating lobes. A split into two arrays, sharing the back-end, reduces these artifacts significantly and possibly increases noise and antenna performance e.g. by reducing sky noise at the highest frequency.

Various types of antenna elements are being investigated emphasizing the potential for excellent cost-performance ratios. For example, bow-tie and log-periodic antenna elements and array prototypes are being developed because of their potential capabilities to cover the entire AA-low band and a 5 times frequency scaled (350–2,250 MHz) model of the conical log spiral antenna is being simulated and prototyped as an example of a wideband band antenna for array development all offering potentially very low cost.

In the final SKA_1, a system consisting of 50 stations, each 180 m diameter and 11,200 antennas, can fulfill the above sensitivity requirement by creating a total physical collecting area of 1.3 km^2.

The required bandwidth for AA-low implies an instantaneous bandwidth of 380 MHz, but it is preferable due to RF-effects (such as band-pass roll-off) to over-sample the incoming signal at 1 GS/s (500 MHz instantaneous B/W). Whilst the exact bit-depth of the Analogue-to-Digital Converter depends on many factors including the strength of interferers as well as the properties of the RF signal path, we take the conservative number of 8-bits, affording us about 48 dB dynamic range. After the signal has been suitably digitized, the signal processing functionality naturally falls into three main areas: Channelization, beamforming and correlation. Putting the correlation aside for the moment, it is expected that both the spectral and the spatial decomposition of the incoming bandwidth and the FoV will follow a hierarchical structure.

A dense AA station for the SKA has of the order of 75,000 dual polarization receiver chains. As for AA-low this implies a large amount of electronics which has severe cost and power implications. The dense array has effectively a fixed collecting area, so, is used where the Sky noise is relatively low and constant above ~400 MHz. Hence, the array sensitivity will be dominated by receiver noise giving the requirement for the lowest possible front-end noise performance for good sensitivity. The top frequency is limited by the number of elements that can be afforded; to stretch to 1.4 GHz we have allowed the system to become gradually sparse above 1 GHz

An important decision for the array system design is how many elements in a cluster are beamformed using analogue techniques prior to using digital processing. Analogue beamforming is cheaper today than a digital system, but has some significant limitations: each beam that is formed requires another set of hardware; it is hard to have precise calibration, particularly to correct polarization issues; and analogue systems have potential drift issues. Digital signal processors, DSP, can implement high precision calibration and beamforming, and can provide a large number of beams just by using additional processing and communications capability. Analogue beamforming could use true time delays, TTDs, e.g. using circuit board tracks, but these are large; alternatively, phase shifters can be used, since they are easier to integrate, but they are relatively narrow band because low frequencies are delayed more than high frequencies, preventing a full bandwidth coherent beam. The ideal architecture is to digitize every element path and perform all the beamforming in the digital domain. This is very flexible; however, it is currently more expensive and higher power than an analogue system. The expectation is for a 2018 implementation to use TTD for small clusters of elements, probably four, and digitize the single, very large, beam that is produced. DSPs then form the very large number of beams from the array, within a field of view defined by the analogue beam.

The maximum instantaneous bandwidth of the system is the full frequency range that the array operates over. There may be some restrictions, discussed above; however the more important consideration is the total data rate from the array, which ultimately defines the performance of the system. With a flexible system, which is ideal for maximizing the science output, beams will not necessarily need to form conventional beams, but can be tailored over the observed sky as a function of frequency; for example to cover a relatively narrow frequency band over a larger observed sky area e.g. for transient event search; or create a constant field of view independent of frequency.

The sensitivity of the system is a function of frequency and is determined from: size and number of arrays, system temperature (T_{sys}), scan angle and the apodisation employed. A critical factor is the receiver noise, dominated by the front-end: element, first amplifier and their matching. Due to the number of receivers and the physical size of the array it is not practical to consider cooling the front-ends for improved noise figures. Consequently, the system will be running at ambient temperature. Progress in this area has been significant and the current best front-ends have a T_{sys} of <60 K (including sky and receiver noise), with <50 K expected in 2011. The required <40 K is expected to be achieved before 2016.

Because an AA is essentially a major processing system with receiver inputs, the dynamic range and polarization purity requirements of the SKA are achievable, but will require sufficient analogue stability, and use an array of sufficient diameter to measure and counter the atmospheric effects. Only the required number of arrays for imaging quality will be deployed, the central processing requirements are thus reduced to a reasonable level considering the survey speeds and data rates that are achieved. In effect the processing in the arrays is mitigating the central processing load.

4 Technology Roadmap

The technologies required for the AAs are generally the focus of the ICT industry: faster and lower power processing, higher speed communications, lower cost, increased storage etc. Hence, AAs tend over time to get more practical, with better performance. At some time, dense AAs become affordable and indeed cheaper than traditional dishes. By carefully studying technology roadmaps for fabrication capabilities, discussion with major semiconductor and communication companies and review of markets that require similar components we anticipate that a dense AA system for the SKA meeting performance, cost and power constraints can be scheduled for 2018 construction.

The key components for the array itself, not currently available, are: an LNA with <15 K noise and an analogue system with <150 mW total power; a DSP, chip providing >20TMACs and use ~25 watts integrated with up to 128 6-bit 3 GS/s digitizers each using <100 mW; a programmable DSP of >20TMACS with 128 I/Os of >10 Gb/s each; and short-range ~50 m pluggable optical links of >120 Gb/s with >2.5 W power. There is a similar requirement list for the central processing systems. These are all projected and indeed will improve post 2018.

5 Conclusions

An outline of the required effort and initial developments have been discussed which should lead to a production-ready design of the low frequency aperture array component of SKA_1. Significant effort will be required but assessments indicate good possibilities for achieving the SKA_1 specifications. This will be supported by science simulations vis a vis the design parameters e.g. with respect to wide field polarimetric capabilities. If dense AAs can be implemented within the cost and power constraints in the timescale of the SKA they are the most capable technology available, representing almost the perfect collector system. The research performed in the SKADS program shows that technology evolution will be sufficient to enable substantial deployment in the second phase of the SKA starting in 2018.

Acknowledgements This paper is due to the work of many people in the SKADS project and continuing into the AAVP. The authors do not wish to list just a few participants, but to fully acknowledge the contributions made by everyone throughout these exciting developments.

References

1. van Ardenne, A., Bregman, J.D., van Cappellen, W.A., Kant, G.W., bij de Vaate, J.G.: "Extending the field of view with phased array techniques". Proc. IEEE **97**(8), (2009)
2. Dewdney, P.E., Hall, P.J., Schilizzi, R.T., Lazio, T.J.L.W.: "The square kilometre array". Proc. IEEE **97**(8), (2009). www.skatelescope.org
3. "Square Kilometre Array Design Studies, SKADS", www.skads-eu.org
4. Torchinsky, S. et al.: Proc. SKADS Conference, Limelette, Oct. 2009, pp 9–14, ISBN 978-90-805434-5-4, March 2010
5. Faulkner, A.J. et al.: "SKA Memo 122: Aperture Arrays for the SKA – the SKADS White Paper", 2010
6. Garrett, M.A. et al.: "SKA Memo 125: A Concept Design for SKA Phase 1 (SKA1)", 2010
7. de Vos, M., Gunst, A.W., Nijboer, R.: "The LOFAR telescope: system architecture and signal processing". Proc. IEEE **97**(8), (2009)
8. Lonsdale, C.J. et al.: "The Murchison widefield array: design overview". Proc. IEEE **97**(8), (2009)
9. "Aperture Array Verification Program", www.ska-aavp.eu
10. Lazio, T.J.L.W.: "Design reference mission: SKA Phase1", Phase1-DRM-V1.3, www.skatelescope.org, 2011

AGN, Star Formation, and the NanoJy Sky

Paolo Padovani

Abstract I present simple but robust estimates of the types of sources making up the faint, sub-μJy radio sky. These include star-forming galaxies and radio-quiet active galactic nuclei but also two "new" populations, that is low radio power ellipticals and dwarf galaxies, the latter likely constituting the most numerous component of the radio sky. I then estimate for the first time the X-ray, optical, and mid-infrared fluxes these objects are likely to have, which are very important for source identification and the synergy between the upcoming SKA and its various pathfinders with future missions in other bands. On large areas of the sky the SKA, and any other radio telescope producing surveys down to at least the μJy level, will go deeper than all currently planned (and past) sky surveys, with the possible exception of the optical ones from PAN-STARRS and the LSST. On the other hand, most sources from currently planned all-sky surveys, with the likely exception of the optical ones, will have a radio counterpart within the reach of the SKA. JWST and the ELTs might turn out to be the main, or perhaps even the only, facilities capable of securing optical counterparts and especially redshifts of μJy radio sources.

1 Introduction

The radio bright ($\gtrsim$1 mJy) sky consists for the most part of active galactic nuclei (AGN) whose radio emission is generated from the gravitational potential associated with a supermassive black-hole and includes the classical extended jet and double lobe radio sources as well as compact radio components more directly associated with the energy generation and collimation near the central engine. Below 1 mJy there is an increasing contribution to the radio source population from synchrotron emission resulting from relativistic plasma ejected from supernovae associated with

P. Padovani (✉)
ESO, Garching bei München, Germany
e-mail: ppadovan@eso.org

D. Barbosa et al. (eds.), *The Square Kilometre Array: Paving the way for the new 21st century radio astronomy paradigm*, Astrophysics and Space Science Proceedings, DOI 10.1007/978-3-642-22795-0_3,

massive star formation in galaxies. After years of intense debate, however, this contribution appears not to be overwhelming, at least down to $\sim$50 μJy. Deep ($S_{1.4\,\mathrm{GHz}} \geq 42\,\mu$Jy) radio observations of the VLA-Chandra Deep Field South (CDFS), complemented by a variety of data at other frequencies, imply a roughly 50/50 split between star-forming galaxies (SFG) and AGN [15], in broad agreement with other recent papers [22, 23]. About half of the AGN are radio-quiet, that is of the type normally found in optically selected samples and characterised by relatively low radio-to-optical flux density ratios and radio powers [15], which constitute an almost negligible minority above 1 mJy.

This source population issue is strongly related to the very broad and complex relationship between star formation and AGN in the Universe. Although the details are still not entirely settled, there is in fact increasing evidence that in the co-evolution of supermassive black holes and galaxies nuclear activity plays a major role through the so-called "AGN Feedback" (e.g., [4]). Radio observations afford a view of the Universe unaffected by the absorption, which plagues most other wavelengths, and therefore provide a vital contribution to our understanding of this co-evolution. However, while we have a reasonably good handle on the radio evolution and luminosity functions (LFs) of powerful sources (e.g., radio quasars), the situation for the intrinsically fainter, and therefore more numerous, radio sources is still murky. Moreover, there are still no published results on the radio evolution of radio-quiet AGN, which are intrinsically weak sources ($P_{1.4\,\mathrm{GHz}} \lesssim 10^{24}$ W Hz^{-1}) and a non-negligible component of the sub-mJy sky. Deeper radio observations over large areas of the sky are then desperately needed to determine the LF and evolution of the most common radio sources in the Universe.

These will soon be realised, as radio astronomy is at the verge of a revolution, which will usher in an era of large area surveys reaching flux density limits well below current ones. The Square Kilometre Array (SKA),[1] in fact, will offer an observing window between 70 MHz and 10 GHz extending well into the *nanoJy* regime with unprecedented versatility. The field of view will be large, up to $\sim$200 deg^2 below 0.3 GHz and possibly reaching $\sim$25 deg^2 at 1.4 GHz. First science with $\sim$10% SKA should be near the end of this decade. Location will be in the southern hemisphere, either Australia or South Africa. Many surveys are being planned with the SKA, possibly comprising an "all-sky" 1 μJy survey at 1.4 GHz and an HI survey out to redshift $\sim$1.5, which should include $\sim$10^9 galaxies.

The SKA will not be the only participant to this revolution. The LOw Frequency ARray (LOFAR),[2] the Expanded Very Large Array (EVLA),[3] the Australian Square Kilometre Array Pathfinder (ASKAP),[4] the Allen Telescope Array,[5] Apertif,[6]

[1] http://www.skatelescope.org.

[2] http://www.astron.nl/radio-observatory/astronomers/lofar-astronomers.

[3] http://science.nrao.edu/evla.

[4] http://www.atnf.csiro.au/projects/askap/.

[5] http://ral.berkeley.edu/ata.

[6] http://www.astron.nl/general/apertif/apertif.

Meerkat,[7] and others, will survey the sky vastly faster than is possible with existing radio telescopes producing surveys covering large areas of the sky down to fainter flux densities than presently available.

What lies beneath the surface of the deepest surveys we currently have, which reach $\approx 10\,\mu$Jy at a few GHz? Predictions for the source population at radio flux densities $<1\,\mu$Jy have been made, amongst others, by [24]. These authors have presented detailed estimates for the number counts of faint radio sources, which show, for example, that SFG should make up $\sim$90% of the total population at $S_{1.4\,\mathrm{GHz}} \sim 1\,\mu$Jy but which had to rely, for obvious reasons, on extrapolations. Moreover, radio-quiet AGN were included in the modelling by converting their X-ray LF to the radio band assuming a linear correlation between radio and X-ray powers. Most importantly, as described below, two crucial constituents of the sub-μJy sky have been excluded by all previous studies.

2 MicroJy and NanoJy Radio Source Population

I present here a simple approach to study the radio sky source population, based on only two parameters: the smallest flux density and the largest surface density of radio sources. The main idea is to provide robust results based on some basic observables and to pay particular attention to *all* populations reaching below the μJy level.

The smallest flux density f_{lim} of a population of sources depends on the minimum radio power at $z \sim 0$, $P_{\mathrm{min}}(0)$, the maximum redshift of the sources, z_{max}, and any luminosity evolution $le(z)$, where $P(z) = P(0) \times le(z)$. If evolution peaks at z_{top} ($\leq z_{\mathrm{max}}$) and then stops, then $f_{\mathrm{lim}} = P_{\mathrm{min}}(0) le(z_{\mathrm{top}})(1 + z_{\mathrm{max}})^{1-\alpha}/4\pi D_L^2(z_{\mathrm{max}})$[8] where $D_L(z)$ is the luminosity distance. Luminosity evolution makes sources brighter and therefore increases f_{lim}.

The largest surface density of a population, $N(>f_{\mathrm{lim}})$, depends on its number density at $z \sim 0$, $N_{\mathrm{T}}(0)$, the maximum redshift z_{max}, and any density evolution $de(z)$, where $N_{\mathrm{T}}(z) = N_{\mathrm{T}}(0) \times de(z)$. Then $N(> f_{\mathrm{lim}}) = (N_{\mathrm{T}}(0)/4\pi) \int_0^{z_{\mathrm{max}}} de(z) dV/dz$ sr^{-1} where dV/dz is the derivative of the comoving volume. In case of no density evolution $N(>f_{\mathrm{lim}}) = N_{\mathrm{T}}(0) V(z_{\mathrm{max}})/4\pi$ sr^{-1}.

Information on the local LF, needed to derive $N_{\mathrm{T}}(0)$ and $P_{\mathrm{min}}(0)$, and the evolution for various classes was derived from a variety of sources (see [14]). The resulting flux and surface density limits are shown in Fig. 1. These values should be considered as robust upper and lower limits respectively, because: (1) one cannot exclude that lower-power, and therefore more numerous, objects exist; (2) z_{max} could be larger than assumed ($\sim$6 for the majority of classes).

[7] http://www.ska.ac.za/meerkat.

[8] Throughout this paper spectral indices are written $S_\nu \propto \nu^{-\alpha}$, magnitudes are in the AB system, and the values $H_0 = 70$ km s^{-1} Mpc^{-1}, $\Omega_{\mathrm{M}} = 0.3$, and $\Omega_\Lambda = 0.7$ have been used.

Figure 1 shows that the most powerful radio sources are, not surprisingly, the ones having the largest flux density ($\approx 0.1 - 1$ mJy) and the smallest surface density ($\approx 1 - 50$ deg^{-2}) limits. BL Lacs are only slightly fainter then Fanaroff–Riley (FR) IIs, while FR Is are the only radio-loud sources reaching $\approx 1\,\mu$Jy. Radio-quiet AGN and SFG are the faintest classes, going into the nanoJy regime, with SFG dominating the faint radio sky (amongst "classical" radio sources: see below).

I argue that two other populations play a major role at $S_{1.4\,\mathrm{GHz}} < 1\,\mu$Jy. The first one is that of low-power ellipticals. It has been know for quite some time that ellipticals of similar optical luminosity vary widely in radio power [17]. Miller et al. [13] have studied the radio LF in the Coma cluster and found that 58% of red sequence galaxies with $M_r \leq -20.5$ are undetected at about 28 μJy r.m.s. Stacking these sources, they obtained a detection corresponding to $P_{1.4\,\mathrm{GHz}} \sim 3 \times 10^{19}\,\mathrm{W\,Hz^{-1}}$. These faint ellipticals are *not* represented in previous models of the sub-μJy sky: for example, the lower limit of the radio-loud AGN LF in [24] is equivalent to $P_{1.4\,\mathrm{GHz}} \sim 2\times10^{20}\,\mathrm{W\,Hz^{-1}}$. For $P_{\mathrm{min}} < 3\times10^{19}\,\mathrm{W\,Hz^{-1}}$, assuming no luminosity evolution, moderate negative density evolution, and taking the number density of all early-type galaxies I get $f_{\mathrm{lim}} < 0.6$ nanoJy and a limiting surface

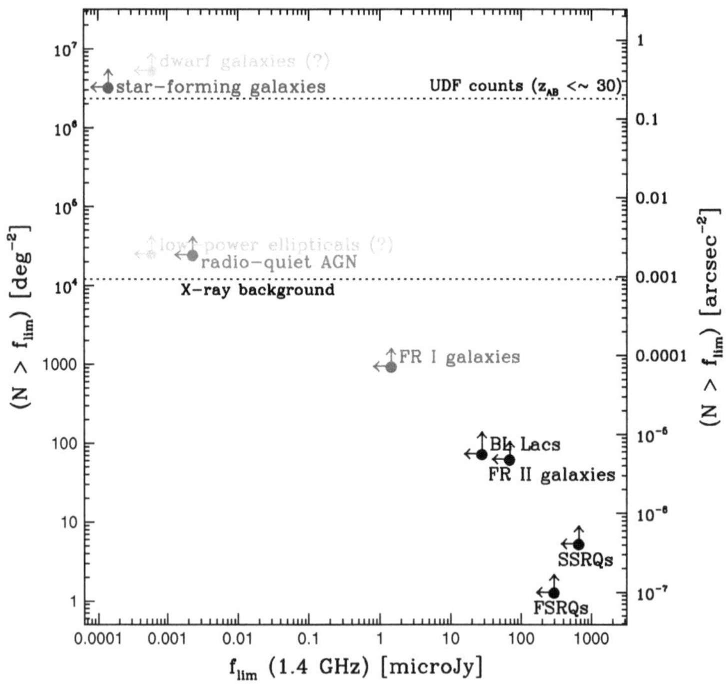

Fig. 1 The largest surface density vs. the smallest flux density for various classes of radio sources (FSRQs = flat-spectrum radio quasars, SSRQs = steep-spectrum radio quasars, FR I and FR II = Fanaroff–Riley type I and II radio galaxies). The two *horizontal lines* denote, from *top* to *bottom*, the surface density of the optical sources in the Hubble Ultra Deep Field and the surface density of the AGN needed to explain the X-ray background

density for low-power ellipticals $\approx 2.4 \times 10^4\,\mathrm{deg}^{-2}$, of the same order as that of radio-quiet AGN.

The other population missing from previous studies is that of dwarf galaxies, which are very faint and constitute the most numerous extragalactic population. This class includes dwarf spheroidals and ellipticals, dwarf irregulars, and blue compact dwarf galaxies (BCDs), and it has never been considered for the simple reason that its radio LF and evolution has never been determined. But the simple approach adopted here can provide us with some idea of how faint and how numerous these sources are going to be in the sub-μJy sky. Since most galaxies at the faint end of the LF are blue (e.g., [2, 20]), most dwarfs in the Universe should be of the star-forming type and I then assume the same luminosity evolution as for SFG. For $P_{\mathrm{min},1.4\,\mathrm{GHz}} < 1.6 \times 10^{18}\,\mathrm{W\,Hz}^{-1}$ [9] and taking also the faint end of the Sloan Digital Sky Survey (SDSS) LF [2] I derive $f_{\mathrm{lim}} < 0.6$ nanoJy and a (likely) conservative limiting surface density $\approx 5 \times 10^6\,\mathrm{deg}^{-2}$, higher than all other classes. *Dwarf galaxies are then likely to be the most numerous component of the faint radio sky.*

3 Multi-Wavelength Properties of MicroJy and NanoJy Radio Sources

Detecting sources is only part of the story, as then comes the identification process. This requires a wealth of multi-wavelength data, ranging from the optical/near-IR imaging needed to provide an optical counterpart and, when needed, photometric redshifts, to the optical/near-IR spectra required to estimate a redshift, and hence the distance of sources, to the X-ray data, which are vital to separate AGN from SFG (e.g., [15] and references therein), to the mid-infrared colours, which provide additional information on this separation (e.g., [19]). I here provide estimates for the the X-ray, optical, and mid-infrared fluxes these sources are likely to have. Optical magnitudes will also determine how feasible it will be to obtain redshifts for them. This kind of information is important for planning purposes, to be ready to take full advantage of the new, deep radio data, and also to maximise the synergy between the SKA and its pathfinders, and present but also, most importantly, future missions. To the best of my knowledge, these estimates have never been made before.

In the following I will assume that low-power ellipticals have the same multi-wavelength properties as their higher-power relatives, with the obvious caveat that, for the same galaxy optical magnitude they will have a much lower radio emission, and therefore will be characterised by a much lower radio-to-optical flux density ratio. As regards dwarf galaxies, my working hypothesis, which is not contradicted by the data, will be that dwarf spheroidals and ellipticals have spectral energy distributions similar to those of low-power ellipticals, while dwarf irregulars and BCDs are mini-spirals. As mentioned above, the faint radio sky should be dominated by the star-forming type of dwarfs.

3.1 X-ray Band

Figure 2 plots 0.5–2 keV flux vs. 1.4 GHz radio flux density and shows the loci of X-ray selected and radio-selected, radio-quiet AGN, SFG, and FR Is. Note that radio-quiet AGN will span the full range between the two dashed lines in Fig. 2, with radio (X-ray) selection favouring sources with low (high) X-ray-to-radio flux density ratios. The position of these loci with respect to survey limits determines the fraction of sources of a given class detected in one band with counterparts in the other. The figure shows also the expected X-ray fluxes for "typical" radio-quiet AGN, SFG, and FR Is (from the NASA/IPAC Extragalactic Database [NED]) scaled to 1 μJy. The mean radio and X-ray flux values, or upper limits, for sources

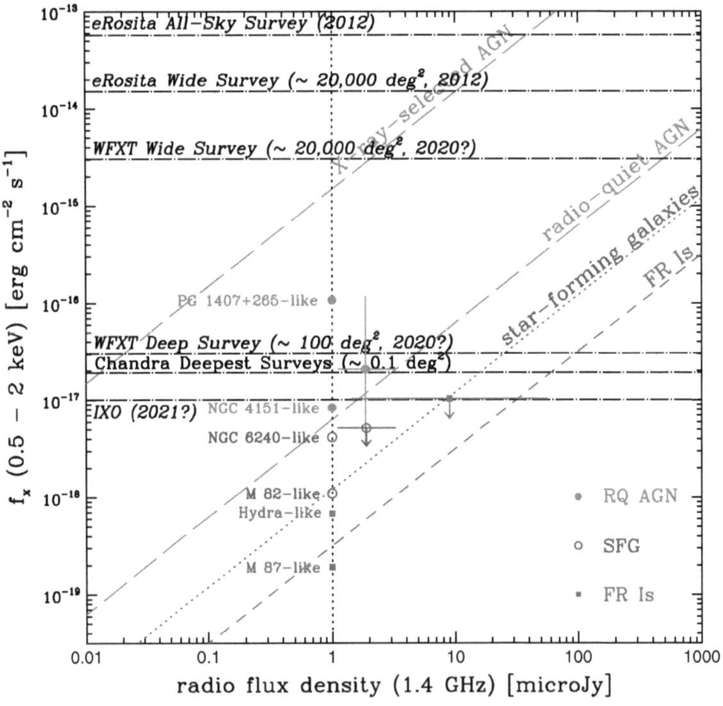

Fig. 2 0.5–2 keV X-ray flux vs. the 1.4 GHz radio flux density for faint radio sources. The loci of X-ray selected and radio-selected, radio-quiet AGN (*long-dashed lines*), SFG (*dotted line*), and FR Is (*short-dashed line*) are indicated. The scaled X-ray fluxes of prototypical representatives of the three classes at $S_{1.4\,\mathrm{GHz}} = 1\,\mu$Jy are also shown. Finally, the mean radio and X-ray flux values, or upper limits, for sources belonging to an hypothetical sample characterised by $S_{1.4\,\mathrm{GHz}} \geq 1\,\mu$Jy, as extrapolated from the VLA-CDFS sample, with *error bars* indicating the standard deviation, are also marked. The *horizontal dot-dashed lines* indicate the limits of (from *top* to *bottom*): the eRosita All-Sky and Wide Surveys, the WFXT Wide and Deep Surveys, Chandra's deepest surveys, and IXO. Survey areas and launch dates for future missions, or best guesses at the time of writing, are also shown

belonging to an hypothetical sample characterised by $S_{1.4\,\mathrm{GHz}} \geq 1\,\mu$Jy, derived from the VLA-CDFS sample [15] are also shown. The fact that the three estimates give consistent results is reassuring and shows that we can predict reasonably well the X-ray fluxes of faint radio sources.

The horizontal dot-dashed line at $f_{0.5-2\,\mathrm{keV}} \sim 2 \times 10^{-17}\,\mathrm{erg\,cm^{-2}\,s^{-1}}$ represents the deepest X-ray data currently available, that is the Chandra Deep Field South 2 Ms Survey [10], covering about 0.1 deg^2. The other horizontal lines indicate, from top to bottom, the limiting fluxes for point sources for surveys to be carried out with eRosita[9] and the Wide Field X-ray Telescope[10] (WFXT). The faintest X-ray flux limit corresponds to the International X-ray Observatory (IXO),[11] which will provide the deepest X-ray view on the Universe for quite some time. Like Chandra, IXO will not produce large area surveys.

The main message of Fig. 2 is that even the most powerful X-ray missions we are going to have for the next 20 years or so will only detect the counterparts of radio-quiet AGN with radio flux densities down to $\approx$1 μJy. The bulk of the μJy population, which is most likely to be made up of SFG, will have X-ray fluxes beyond even the reach of IXO. Same, or possibly even worse, story for FR Is. The situation will obviously be even more critical in the nanoJy regime, where very few radio sources will have an X-ray counterpart in the foreseeable future.

On the positive side, basically all extragalactic sources in the eRosita All-Sky and Wide Surveys and WFXT Wide Survey will have an SKA counterpart, as they will have $S_{1.4\,\mathrm{GHz}} > 1\,\mu$Jy. This should help in the identification work of, for example, the ten million or so point sources expected in the latter, by also providing very accurate positions.

3.2 *Optical Band*

At variance with the X-ray case, there is in general no apparent correlation between radio and optical powers in extragalactic sources. This is due to the fact that while the radio, far-infrared, and X-ray bands all trace the star formation rate (SFR) (e.g., [7,16]), the optical band does not, as young stars dominate the ultraviolet continuum. Moreover, as mentioned in Sect. 2, elliptical galaxies of similar optical power can host radio sources differing by huge amounts in their radio luminosity.

However, SFG and radio-quiet AGN can only reach a reasonably well-defined value of the K-corrected radio-to-optical flux density ratio $R = \log(S_{1.4\,\mathrm{GHz}}/S_{\mathrm{R_{mag}}})$, where $S_{\mathrm{R_{mag}}}$ is the R-band flux density (e.g., [11, 15]). Converting from the values derived for the two classes in the V-band by [15] and taking the average one gets for the R-band a maximum value $R \approx 1.4$. Due to K-correction effects, *observed* R values for SFG and radio-quiet AGN can be >1.4. It has to be noticed that, while

[9] http://www.mpe.mpg.de/heg/www/Projects/EROSITA/main.html.

[10] http://wfxt.pha.jhu.edu/.

[11] http://ixo.gsfc.nasa.gov/.

all "classical" radio-loud quasars have $R > 1.4$, this is not the case for many radio-galaxies (e.g., [15]), which can extend to $R < 1.4$. Indeed, as mentioned above, low-power ellipticals will have lower R values than FR Is.

The absolute SFR in star-forming galaxies spans a very large range, from $\sim 20\ M_\odot\ \mathrm{year}^{-1}$ in gas-rich spirals to $\sim 100\ M_\odot\ \mathrm{year}^{-1}$ in optically selected starburst galaxies and up to $\sim 1{,}000\ M_\odot\ \mathrm{year}^{-1}$ in the most luminous IR starbursts [7]. Therefore, for the same optical magnitude "normal" spirals will be characterised by lower radio emission than starbursts, and therefore will have a smaller radio-to-optical flux density ratio ($R \lesssim 0.1 - 0.3$).

Dwarf galaxies also have low flux density ratios. From the mean values given by [9] I get $R \approx -0.6$ (converting to my notation), while Fig. 9 of [8] shows that $R < -0.8$ for $M_B \sim -17$. Even BCDs, which are the most star-forming amongst dwarfs, are characterised by $R \approx -0.2$ [6].

Figure 3 plots R_{mag} vs. 1.4 GHz radio flux density and shows: (1) the expected R_{mag} for "typical" radio-quiet AGN, SFG (starbursts, spirals, and dwarfs), and FR Is (from NED) scaled to 1 μJy; (2) the maximum value for SFG and the approximate dividing line between radio-loud and radio-quiet AGN (diagonal dashed line). SFG and radio-quiet AGN are expected to populate the top left part of the diagram; (3) the mean radio and R_{mag} values for sources belonging to an hypothetical sample characterised by $S_{1.4\,\mathrm{GHz}} \geq 1\ \mu$Jy, based on VLA-CDFS data. The three methods give consistent results, which is reassuring and shows that we have a reasonable handle on the magnitudes of faint radio sources.

The horizontal dot-dashed line at $R_{\mathrm{mag}} \sim 29.3$ represents the limit of the deepest optical data currently available, the Hubble UDF [1], covering about 11 arcmin^2. The other horizontal lines indicate, from top to bottom, the limiting magnitudes for the Panoramic Survey Telescope and Rapid Response System (PAN-STARRS),[12] which from Hawaii will survey about 3/4 of the sky down to $R_{\mathrm{mag}} \sim 26$ during 10 years of operation and the Large Synoptic Survey Telescope (LSST),[13] which will be located in Chile and will provide a survey of about half the sky down to $R_{\mathrm{mag}} \sim 27.5$ during 10 years of operation. Further below there are the limiting magnitudes for point sources for the James Webb Space Telescope (JWST)[14] and for the European Extremely Large Telescope[15] (E–ELT)[16] for the given signal-to-noise (S/N) ratios and exposure times.[17] Both these telescope will operate in observatory mode. It is

---

[12]http://pan-starrs.ifa.hawaii.edu/.

[13]http://www.lsst.org.

[14]http://www.stsci.edu/jwst/.

[15]http://www.eso.org/sci/facilities/eelt/.

[16]I am using the E–ELT as an example because it has the largest mirror (42 m in diameter) amongst the three very large telescopes being planned. The other two are the Thirty Meter Telescope (TMT; http://www.tmt.org/) and the Giant Magellan Telescope (GMT; http://www.gmto.org/).

[17]The JWST limit comes from the JWST Web pages. The E–ELT limit was derived from the E–ELT Exposure Time Calculator (ETC) using the Laser-Tomography/Multi-Conjugate Adaptive Optics option.

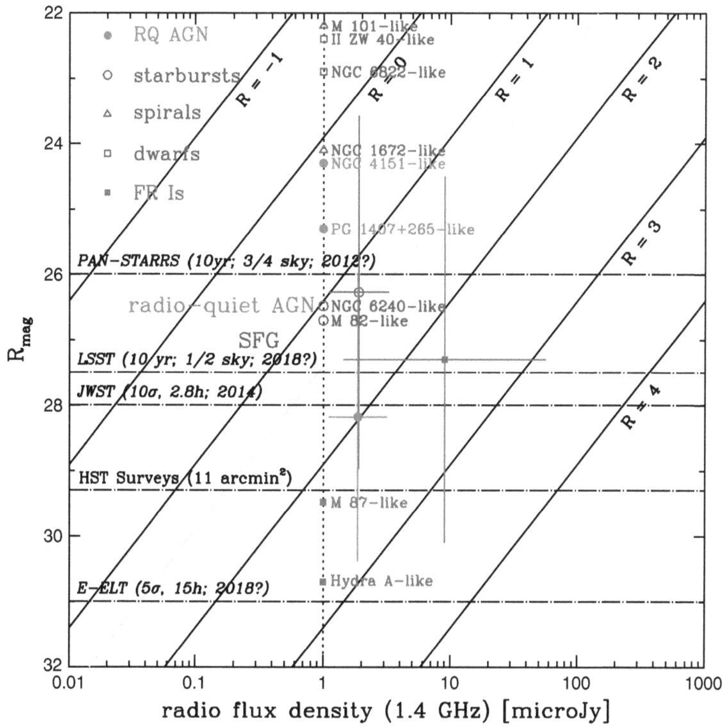

Fig. 3 $R_{\rm mag}$ vs. the 1.4 GHz radio flux density for faint radio sources. *Diagonal lines* represent different values of $R = \log(S_{1.4\,\rm GHz}/S_{\rm R_{mag}})$, ranging from -1 (*top*) to 4 (*bottom*). The *diagonal dashed line* at $R = 1.4$ indicates the maximum value for SFG and the approximate dividing line between radio-loud and radio-quiet AGN, with SFG and radio-quiet AGN expected to populate the *top left* part of the diagram. The scaled R magnitudes of prototypical representatives of the three classes at $S_{1.4\,\rm GHz} = 1\,\mu$Jy are also shown, with SFG split into starbursts, spirals, and dwarfs. Finally, the mean radio and $R_{\rm mag}$ values for sources belonging to an hypothetical sample characterised by $S_{1.4\,\rm GHz} \geq 1\,\mu$Jy, as extrapolated from the VLA-CDFS sample, with *error bars* indicating the standard deviation, are also marked. The *horizontal dot-dashed lines* indicate the approximate point-source limits of (from *top* to *bottom*): PAN-STARRS, LSST, JWST, the deepest Hubble Space Telescope surveys, and the E–ELT. Survey areas (for LSST and PAN-STARSS), S/N ratios and exposure times (for the E–ELT and JWST) and operation/launch dates for future missions, or best guesses at the time of writing, are also shown

important to keep in mind that all of these limits are only approximate, as none of these telescopes are in operation yet.

Given the progression towards lower R values going from (non-dwarf) starbursts, to spirals, and to BCDs and dwarf irregulars, the typical magnitudes of faint radio sources at a given flux density will depend critically on which of these SFG subclasses dominates the counts.

The contribution of "normal" spirals to faint radio number counts is not well known. Defining as a "normal", non-starburst source a galaxy with an SFR

$<10\ M_\odot\,\mathrm{yr}^{-1}$ and converting this to a radio power translates to $P_{1.4\,\mathrm{GHz}} < 8.4 \times 10^{21}\,\mathrm{W\,Hz}^{-1}$. By splitting the SFG LF of [18] into two at $P_{1.4\,\mathrm{GHz}} = 8.4 \times 10^{21}\,\mathrm{W\,Hz}^{-1}$ and evaluating the number counts assuming the evolution of [5] I find that "normal" spirals should be more numerous than starbursts for $S_{1.4\,\mathrm{GHz}} \lesssim 5\,\mu$Jy. Moreover, by making an educated guess on the radio LF of dwarf galaxies by converting their optical LF and assuming the same evolution as SFG I find that dwarf galaxies should become the most numerous constituents of the radio sky at flux densities 20 nanoJy $\lesssim S_{1.4\,\mathrm{GHz}} \lesssim$ 300 nanoJy.

Figure 3 shows that, down to 1 μJy, most starburst-like SFG and radio-quiet AGN should be detected by the LSST, that is they will have a counterpart in a large area survey. To be more specific, based on the VLA-CDFS extrapolation, $\sim$66% of starburst galaxies with $S_{1.4\,\mathrm{GHz}} \geq 1\,\mu$Jy should have $R_{\mathrm{mag}} < 27.5$. This number represents an upper limit because by scaling the VLA-CDFS magnitudes I have assumed that the mean redshift is unchanged, while [15] have found a strong correlation between redshift and magnitude. Therefore, K-correction effects will be larger and observed magnitudes will be fainter. On the other hand, "normal" spirals should outnumber starbursts for $S_{1.4\,\mathrm{GHz}} \lesssim 5\,\mu$Jy. If even distant spirals are characterised by rest-frame $R \lesssim 0.1 - 0.3$, as is the case for local sources, then μJy sources could be $\approx$2 magnitudes brighter than expected in the case of starbursts and therefore within reach of PAN-STARRS and certainly of the LSST. For fainter radio samples optical magnitudes should get fainter, unless dwarf galaxies, with their low R values, become dominant, which would lead yet to another "brightening". For example, for $R \approx -0.5$ and $S_{1.4\,\mathrm{GHz}} \approx 100$ nanoJy, $R_{\mathrm{mag}} \approx 25$.

FR Is have on average larger R values but also higher radio flux densities, so their mean magnitude in an hypothetical $S_{1.4\,\mathrm{GHz}} \geq 1\,\mu$Jy sample is brighter than those for "typical" sources with $S_{1.4\,\mathrm{GHz}} = 1\,\mu$Jy, although the R values are similar. Low-power ellipticals, however, will have smaller R and therefore brighter magnitudes.

Sources having $R_{\mathrm{mag}} > 27.5$ will obviously be within reach of JWST and the ELTs, which however will be covering a relatively small field of view (up to a few arcmin^2). Depending on the actual relative fraction of starbursts, spirals, and dwarfs, these observatories could be the main (only?) facilities to secure optical counterparts of nanoJy radio sources.

To address the question of what fraction of sources in the LSST and PAN-STARRS surveys will have an SKA counterpart requires a knowledge of the radio properties of very faint optical sources, which at present we do not possess. Nevertheless, one can make some educated guesses. The UDF number counts of [1] show that for $z_{\mathrm{AB}} \lesssim 26$ the surface density is $\sim 3 \times 10^5\,\mathrm{deg}^{-2}$, which implies, based on Fig. 1, that the majority of the objects beyond this limit are star-forming systems. SFG in the VLA-CDFS sample have *observed* $R \approx 1.3$, which means $S_{1.4\,\mathrm{GHz}} \approx 3\,\mu$Jy and $\approx$70 nanoJy at $R_{\mathrm{mag}} \sim 26$ and $\sim$27.5 respectively, that is within reach of the SKA. Optical selection however will be biased towards smaller R and therefore fainter radio flux densities. More importantly, spirals and dwarfs, which might be the most numerous sub-classes, with their lower R values will have fainter radio flux densities for a given magnitude. In summary, the bulk of LSST and PAN-STARRS sources *might not* have radio flux densities within reach of the SKA, but at this point in time one cannot be more specific.

3.3 *Mid-Infrared Band*

The infrared and radio emission are strongly and linearly correlated in SFG, defining what is known as the "IR-radio relation" (e.g., [21] and references therein). The most recent determination of the so-called (K-corrected) q_{24} parameter, based on the COSMOS field and using an IR/radio-selected sample to reduce selection biases, is $q_{24} = \log(S_{24\mu m}/S_{1.4\,\mathrm{GHz}}) = 1.26 \pm 0.13$ [21]. BCDs also appear to have q parameters globally consistent with those of SFG [6].

It turns out that most SFG and radio-quiet AGN should be detected at 24 μm by JWST down to $S_{1.4\,\mathrm{GHz}} \sim 1\,\mu$Jy and by the SPace Infrared telescope for Cosmology and Astrophysics (SPICA)[18] down to $\sim$100 nanoJy. Both telescopes will be covering a relatively small field of view (up to $\approx$15 arcmin^2). Only a tiny fraction of μJy and nanoJy SFG and radio quiet AGN will have a counterpart in the all-sky infrared surveys provided by the Wide-field Infrared Survey Explorer (WISE)[19] and AKARI.[20]

On the other hand, all extragalactic sources in the WISE and AKARI all-sky surveys will easily have an SKA counterpart, as they will be characterised by $S_{1.4\,\mathrm{GHz}} \gtrsim 100\,\mu$Jy.

4 Redshifts of Faint Radio Sources

A vital component in the identification of astronomical sources is redshift, which allows powers to be estimated and LFs to be derived. Long exposures ($\sim$10 h) with 8/10 m telescopes can currently secure spectroscopic redshifts *in the case of strong emission lines* down to $R_{\mathrm{mag}} \approx 26$, so this value represents a hard limit. The current limiting magnitude for "classical" photometric redshifts typically reaches $R_{\mathrm{mag}} \approx 24$ for $z \lesssim 1.5$ (e.g., [12]); photometric redshifts for special classes of sources, e.g., $z \sim 7$ drop-outs can be derived with HST down to $Y_{\mathrm{AB}} \approx 28$ (e.g., [3]).

Based on the estimates above, down to 1 μJy, only approximately half of the starburst-like SFG will be within reach of 8/10 m telescopes in terms of obtaining a spectroscopic redshift. To be more specific, based on the VLA-CDFS extrapolation, $\sim$50% of starburst galaxies with $S_{1.4\,\mathrm{GHz}} \geq 1\,\mu$Jy should have $R_{\mathrm{mag}} > 26$ and $\sim$40% will be above $R_{\mathrm{mag}} = 25$, probably a more realistic limit for a spectrum. As discussed above (Sect. 3.2), these fractions are robust upper limits. On the other hand, spiral, and maybe dwarf galaxies, should be more common than starbursts at μJy and nano-Jy levels, respectively, which would lead to brighter magnitudes, thereby making the job of deriving a redshift easier. Future facilities like the E–ELT

---

[18] http://www.ir.isas.jaxa.jp/SPICA/.

[19] http://wise.ssl.berkeley.edu/.

[20] http://www.ir.isas.jaxa.jp/ASTRO-F/.

and JWST will obviously allow the determination of spectroscopic redshifts for fainter ($R_{\rm mag} > 26$) objects, albeit with relatively long exposures. But even they could have problems in the nanoJy regime.

5 Conclusions

My main results can be summarised as follows:

1. The sub-μJy sky should consist of radio-quiet AGN ($\approx 2 \times 10^4$ deg^{-2}) and star-forming galaxies ($\approx 3 \times 10^6$ deg^{-2}), both of which should get to $S_{1.4\,\rm GHz} \approx 0.1 - 1$ nanoJy. In agreement with previous studies, I find that classical, powerful radio sources, that is radio quasars and FR IIs, do not make it to sub-μJy flux densities and reach $S_{1.4\,\rm GHz} \approx 0.1$ mJy, with BL Lacs and FR Is getting to the $\approx 1 - 10\,\mu$Jy level.
2. Two "new" populations, which have not been considered previously, appear to be very relevant: low-power ($P_{1.4\,\rm GHz} < 10^{20}$ W Hz^{-1}) ellipticals and dwarf galaxies. Using the available (scanty) information, the former should reach similar flux and surface densities as radio-quiet AGN, while the latter could easily be the most numerous component of the faint radio sky ($\gtrsim 5 \times 10^6$ deg^{-2}), with flux densities as low as ≈ 1 nanoJy. Since most galaxies at the faint end of the LF are blue, most dwarfs in the Universe (and therefore most radio sources) should be of the star-forming type.
3. The bulk of the μJy population, which is then most probably going to be made up of star-forming galaxies, is likely to have X-ray fluxes beyond the reach of all currently planned X-ray missions, including IXO. The same applies to FR Is. Even IXO will only detect radio-quiet AGN with radio flux densities $\gtrsim 1\,\mu$Jy. The situation will obviously be worse in the nanoJy regime, where very few radio sources will have an X-ray counterpart in the foreseeable future. On the other hand, basically all extragalactic sources in the eRosita All-Sky and Wide Surveys and WFXT Wide Survey will have an SKA counterpart, as they will be characterised by $S_{1.4\,\rm GHz} > 1\,\mu$Jy. This will help in the identification of, for example, the ten million or so point sources expected in the latter, by also providing very accurate positions.
4. Since the radio, far-infrared, and X-ray emission all trace star formation, this implies relatively tight relationships between these three bands, which in turn means that we can reasonably predict the mid-infrared and X-ray properties of star-forming galaxies with faint radio flux densities. The situation is different in the optical band, where evolved stars dominate. Objects characterised by higher star formation rates will have larger radio-to-optical flux density ratios, which means fainter magnitudes for a given radio flux density. The typical magnitudes of the optical counterparts of faint radio sources depend then on which type of star-forming galaxy (starburst, spiral, or dwarf) will be predominant. Moreover,

in the optical band K-correction and intergalactic absorption effects are important and will result in fainter than expected magnitudes. If, as I estimated, "normal" spirals outnumber starbursts for $S_{1.4\,\mathrm{GHz}} \lesssim 5\,\mu$Jy and if distant spirals are also characterised by relatively low rest-frame radio-to-optical flux density ratios, then most μJy sources should be detected by PAN-STARRS and certainly by the LSST. At fainter radio flux densities optical magnitudes should also get fainter, unless dwarf galaxies, with their lower radio-to-optical flux density ratios, become dominant. Depending on the relative fraction of starbursts, spirals, and dwarfs, JWST and especially the ELTs could be the main, or perhaps even the only, facilities capable of securing optical counterparts of nanoJy radio sources. On the other hand, and for the same reasons discussed above, the bulk of PAN-STARRS and LSST sources might not have radio flux densities within reach of the SKA.

5. As regards redshifts, the same complications described above apply, since the optical band is still involved. Within the same uncertainties, then, many of the sources with $S_{1.4\,\mathrm{GHz}} \geq 1\,\mu$Jy might be too faint for 8/10 m telescopes to be able to provide a redshift determination and the situation might get worse at fainter flux densities, unless dwarf galaxies are prevalent. This means that JWST and particularly the ELTs might be the primary facilities to secure redshifts of μJy radio sources. But even they could have problems in the nanoJy regime.

In summary, the SKA and its pathfinders will have a huge impact on a number of open problems in extragalactic astronomy. Apart from the "obvious" study of star-forming galaxies in their various incarnations and "classical" radio sources, these include also less evident ones, ranging from what makes a galaxy radio-loud (through the study of low-power ellipticals), to why most AGN are radio-quiet (by selecting large samples independent of obscuration), to the incidence and evolution of dwarf galaxies (by providing a "cleaner" radio selection, which might by-pass the surface brightness problems of optical samples).

Identifying faint radio sources, however, will not be easy. On large areas of the sky the SKA will be quite alone in the multi-wavelength arena, with the likely exception of the optical band and even there probably only down to $\approx 1\,\mu$Jy. SPICA, JWST, and especially the ELTs will be a match for the SKA but only on small areas and above $0.1 - 1\,\mu$Jy. At fainter flux densities one might have to resort to "radio only" information, that is HI redshifts, size, morphology, spectral index, etc., although I think this will not be sufficient. On the bright side, most sources from currently planned all-sky surveys, with the likely exception of the optical ones, will have an SKA counterpart.

A more extended discussion of the topics discussed in this paper is given by [14].

Acknowledgements The idea for this work came to me while preparing a talk for the SKA 2010 meeting held in Manchester, UK, in March 2010. My thanks to the organisers of that meeting for inviting me. I also thank Michael Hilker, Ken Kellermann, Jochen Liske, Vincenzo Mainieri, Nicola Menci, Steffen Mieske, and Piero Rosati for useful discussions.

References

1. Beckwith, S.V.W., et al.: Astrophys. J. **132**, 1729 (2006)
2. Blanton, M.R., et al.: Astrophys. J. **631**, 208 (2005)
3. Bouwens, R.J., et al.: Astrophys. J. Lett. **709**, L133 (2010)
4. Cattaneo, A., et al.: Nature **460**, 213 (2009)
5. Hopkins, A.M.: Astrophys. J. **615**, 209 (2004)
6. Hunt, L., Bianchi, S., Maiolino, R.: Astron. Astrophys. **434**, 849 (2005)
7. Kennicutt R.C., Jr.: Annu. Rev. Astron. Astrophys. **36**, 189 (1998)
8. Leon, S., et al.: Astron. Astrophys. **485**, 475 (2008)
9. Leroy, A., Bolatto, A.D., Simon, J.D., Blitz, L.: Astrophys. J. **625**, 763 (2005)
10. Luo, B., et al.: Astrophys. J. Suppl. Ser. **179**, 19 (2008)
11. Machalski, J., Condon, J.J.: Astrophys. J. Suppl. Ser. **123**, 41 (1999)
12. Mainieri, V., et al.: Astrophys. J. Suppl. Ser. **179**, 95 (2008)
13. Miller, N.A., et al.: Astron. J. **137**, 4450 (2009)
14. Padovani, P.: Mon. Not. R. Astron. Soc. **411**, 1547 (2011)
15. Padovani, P., et al.: Astrophys. J. **694**, 235 (2009)
16. Ranalli, P., Comastri, A., Setti, G.: Astron. Astrophys. **399**, 39 (2003)
17. Sadler, E.M., Jenkins, C.R., Kotanyi, C.G.: Mon. Not. R. Astron. Soc. **240**, 591 (1989)
18. Sadler, E.M., et al.: Mon. Not. R. Astron. Soc. **329**, 227 (2002)
19. Sajina, A., Lacy, M., Scott, D.: Astrophys. J. **621**, 256 (2005)
20. Salimbeni, S., et al.: Astron. Astrophys. **477**, 763 (2008)
21. Sargent, M.T., et al.: Astrophys. J. Suppl. Ser. **186**, 341 (2010)
22. Seymour, N., et al.: Mon. Not. R. Astron. Soc. **386**, 1695 (2008)
23. Smolčić, V., et al.: Astrophys. J. Suppl. Ser. **177**, 14 (2008)
24. Wilman, R.J., et al.: Mon. Not. R. Astron. Soc. **388**, 1335 (2008)

Using HI Absorption to Trace Outflows from Galaxies and Feeding of AGN

Raffaella Morganti

Abstract Understanding the role of cold gas in the triggering and evolution of active galactic nuclei (AGN) is one of the goals of future cm and mm facilities. HI 21cm in absorption is one powerful diagnostic that can be used to explore these topics and probe the central regions of AGN. This contribution will briefly summarize some of the recent results in this field including the finding of fast, massive outflows of HI gas that may provide the negative feedback required by the galaxy's evolution models to stop the growth of the black hole and the star formation. The requirements needed for the new radio facilities – and in particular SKA – in order to provide a major step forward in the understanding of the distribution and kinematics of the atomic neutral gas close to the AGN will also be discussed.

1 Why Associated HI Absorption Is Interesting

One of the main goals of the next generation of radio telescopes, and in particular SKA, is to trace the most abundant element in the Universe, neutral hydrogen (HI 21cm), to high redshift and connect this to the formation and evolution of galaxies through the cosmic time. Part of this task will be carried out by observing HI 21cm in emission. However, an important and complementary study can be done by using HI 21cm detected in absorption against strong sources. Here I will report on some recent results in this field and the relevance of SKA and SKA Phase 1 for making a major step forward in this topic.

R. Morganti (✉)
ASTRON, Netherlands Institute for Radio Astronomy, Postbus 2, 7990 AA, Dwingeloo, The Netherlands

Kapteyn Astronomical Institute, University of Groningen, P.O. Box 800, 9700 AV Groningen, The Netherlands
e-mail: morganti@astron.nl

D. Barbosa et al. (eds.), *The Square Kilometre Array: Paving the way for the new 21st century radio astronomy paradigm*, Astrophysics and Space Science Proceedings, DOI 10.1007/978-3-642-22795-0_4,

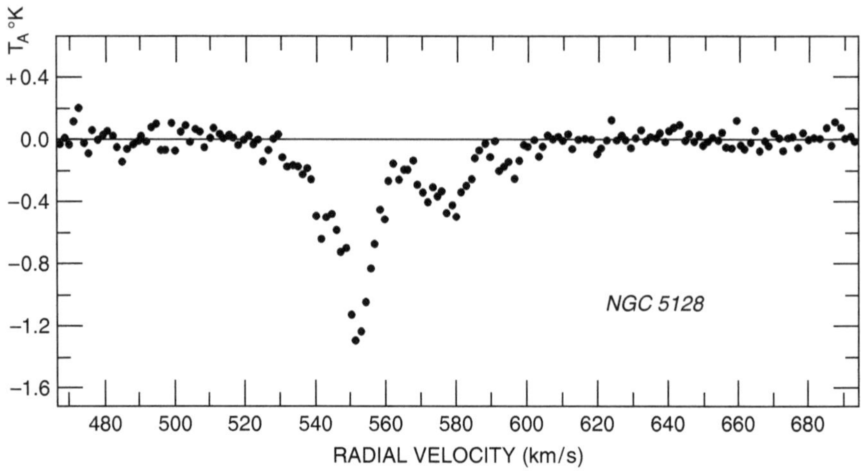

Fig. 1 The first HI absorption detected against the central region of Centaurus A by [25] with the NRAO 140-foot

Neutral hydrogen was detected for the first time in absorption by [25] against the core of the radio source Centaurus A. This result was obtained using the NRAO 140-foot and after that, other detections were obtained with single-dish radiotelescopes. However, the real improvement came with interferometric observations ([14,28,30], to list some initial work). Before describing some of the recent results, it is worth mentioning that there are some advantages and disadvantages in the study of HI in absorption compared to HI emission. The two main advantages are that the detectability of the absorption depends only on the strength of the background continuum, therefore HI absorption can probe the presence of HI at high-z where it is not possible to detect the emission. For the same reason, it can also be detected with high spatial resolution observations, again something impossible for HI emission. This makes the HI absorption indeed ideal for the study of the circumnnuclear regions of AGN. The disadvantage is that it gives a view of the distribution of the HI that is limited to the regions where the background continuum is present. To partly compensate for that, the synergy with other gas diagnostics (i.e. multi-wavebands data) and information about different phases of the gas (ionized and molecular) are crucial (Fig. 1).

The detection of *associated* HI absorption (HI located in the target galaxy) has been identified as a way to probe the *circumnuclear regions of radio-loud AGN*. The typical widths of the absorption profiles are relatively broad ($\geq$100 km/s), i.e. broader than the typical Galactic cold clouds, and this is considered as a signature of the effects of an active black hole.

Thus, unlike *intervening* absorption (HI absorption toward radio-loud background sources, see e.g. [15, 35]), *associated* HI absorption allows not only to understand the HI content of galaxies (as function of their morphological properties, redshift and environment) but also to understand the central region of AGN, their structure and kinematics. This is important because the immediate surroundings of active galactic nuclei are, in fact, complex regions characterized by extreme physical conditions where the interplay between the enormous amounts of energy released from the nucleus and the ISM is critical in determining the evolution of the system.

The role of gas (whether atomic, molecular and ionized) in the nuclear regions is known to be important both in triggering the radio activity and in the subsequent evolution. For example, infalling gas, whether from an accretion disk or an advective flow, is an essential ingredient of AGN activity, as it can feed the central engine and turning a dormant super-massive black hole into an active one. The details of gas accretion (e.g. Bondi accretion from hot gas haloes vs. optically-thick cold accretion disks) can have important implications for the characteristics of the radio emission, producing low-luminosity, edge-darkened FR-I radio galaxies in the first case, and powerful, edge-brightened FR-II systems in the second (e.g. [1]). Conversely, feedback from radio-loud AGN outflows is expected to play a crucial role in regulating star formation in the most massive galaxies (e.g. [7]).

We have now more than fifty radio sources where HI absorption has been detected. Most of the studies of HI absorption have targeted sources stronger than $\sim$100 mJy due to the sensitivity limit of our current radio telescopes. The detection rate is about 15–20% for extended radio galaxies ([9, 13, 18] and references therein) and increases to 30–40% [31] for Compact Steep Spectrum (CSS) and GigaHertz Peaked Sources (GPS), i.e. radio sources supposed to be in the earlier stages of their evolution (see e.g. [10]) radio sources. Thus, there appears to be a tendency for the compact/young sources to be more often detected in HI, suggesting that they are embedded in a richer ISM as expected if a gas-rich merger has triggered the nuclear activity. However, this can also be the result of an observational bias: CSS/GPS sources tend to be strong in radio flux and, therefore, it is more likely to detect absorption in these objects even at low optical depth. An other bias is due to the fact that most of the extended radio sources where HI absorption has been detected so far are low-luminosity, edge-darkened FR-I radio galaxies (typically having relatively strong radio cores and therefore more suitable for observations).

All these observational biases strongly hamper what we can learn from HI absorption observations. In order to improve on this, larger and uniformly selected samples are needed and this is one of the aims of the surveys already planned with the new generation of radio telescopes (from the pathfinders/precursors to SKA) and in particular those that will provide a large instantaneous field of view, therefore, allowing large surveys of the sky.

Finally, a relatively small sub-set of the sources detected in HI 21 cm absorption has been studied in detail using high-resolution VLBI observations. These are key observations for the interpretation of the absorbing screen by tracing the distribution and kinematics of the gas. Complementary data at other wavebands are also key for the interpretation. The general conclusion of these studies is that a variety of structures and phenomena can produce HI absorption.

2 Origin of the HI 21cm Absorption

HI 21cm absorption studies offer the unique possibility of tracing the kinematics and morphology of neutral gas in radio-loud AGN environments, and determining the nature of the accretion or the energetic of the outflow. As mentioned above, a variety of structures can produce HI absorption and often it is quite difficult to disentangle the contribution of these different components.

Circumnuclear tori and disks have been often proposed to produce the HI absorption observed in some radio galaxies but most of the evidence are in fact indirect. In these cases, the absorption is relative broad (typically 100–200 km/s) and centered on the systemic velocity.

The best cases for possible tori (size from pc to tens of pc) have been found in small objects (GPS, CSO etc.) where the background continuum is more favorable for detecting these structures. One of the interesting cases is 4C31.04 [6] where the disk-like structure was confirmed by HCO+ observations from PdB [11]. Other examples can be seen in ([2], and references therein). The high spatial resolution imaging obtained in these studies is only possible in the HI 21cm absorption line observations, due to the presence of background radio emission from relativistic jets on 100 pc scales.

Other cases of circumnuclear disks have been suggested in extended radio galaxies, mainly in low-luminosity FRI type-I and in these galaxies the presence of HI seems to be related to the presence of dust in the circumnuclear regions (van Bemmel et al. in preparation).

Although tracing the HI was hoped to give insight on the fuelling of the AGN (see e.g. [30]), it is clear now that the situation is more complex and it is not always easy to identify infalling gas.

Interesting is the case of Centaurus A where it was claimed evidence of gas infall into the AGN, based on the presence of a redshifted wing in the absorption profile. Deeper observations have recently shown that absorption is detected against the radio core also at velocities blueshifted with respect to the systemic velocity [21]. Moreover, the data show that the nuclear redshifted absorption component is broader than reported before. With these new results, the kinematics of the HI in the inner regions of Cen A appears very similar to that observed in emission for the molecular circumnuclear disk (see Fig. 2), thus likely explained with a cold, circumnuclear disk. On the other hand, the case of NGC 315 [22] indicates the importance of follow up observations at high spatial resolution. In this object, VLBI observations have revealed that part of the HI is concentrated in a cloud located at about 9 pc from the core and in the process of falling into the nucleus. This object appears one of the few cases where possible indications of HI fuelling the AGN have been found.

One of the few examples of possible detection of circumnuclear torus in a powerful extended radio galaxies, has been found in Cygnus A [29], see Fig. 3. However, also in this object not all the HI appears to be located in the disk. Part of the gas has velocities redshifted compared to the systemic suggesting that it is in the process of falling into the nucleus (see also [3]).

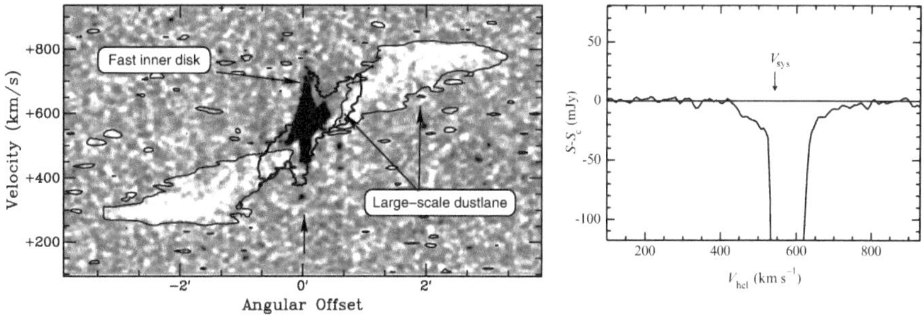

Fig. 2 Position-velocity plot (*left*) of the HI with superimposed the CO emission (thick contour, from [17]) taken along a position angle close to the major axis of the dust lane (see [21] for details). HI absorption is dark and HI emission light colours. The HI absorption profile (*right*) against the central radio core (indicated by the *arrow* in the *left panel*) showing the blue- and redshifted wings of the absorption profile

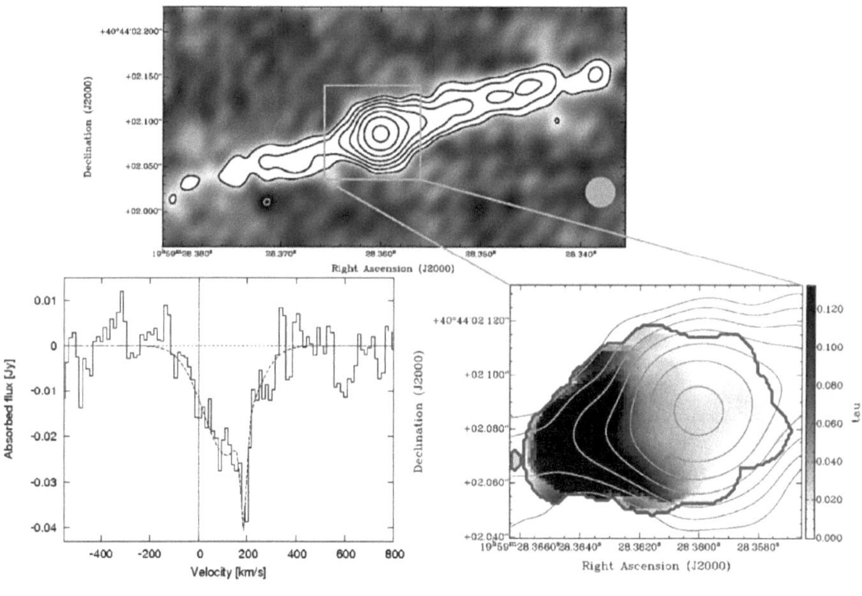

Fig. 3 Panel (from [29]) showing (*top*) the VLBI continuum image at 1.3 GHz of Cygnus A and (*bottom*) the integrated absorption spectrum (left) and the mean opacity (grey scale) over the rest frame velocity range $-80/+170$ km/s

Very intriguing has been the discovery of broad, blue-shifted HI 21cm absorption in a number of nearby radio galaxies (e.g. [20], see Fig. 4). This indicates the existence of fast (>1,000 km/s), massive ($\sim$50 M $\odot$ yr^{-1}) HI outflows [20]. The mass outflow rates are comparable to those obtained for starburst-driven superwinds in ultra-luminous infrared galaxies (e.g. [32]), indicating that AGN outflows could play a similar role to superwinds in enriching the intergalactic medium with

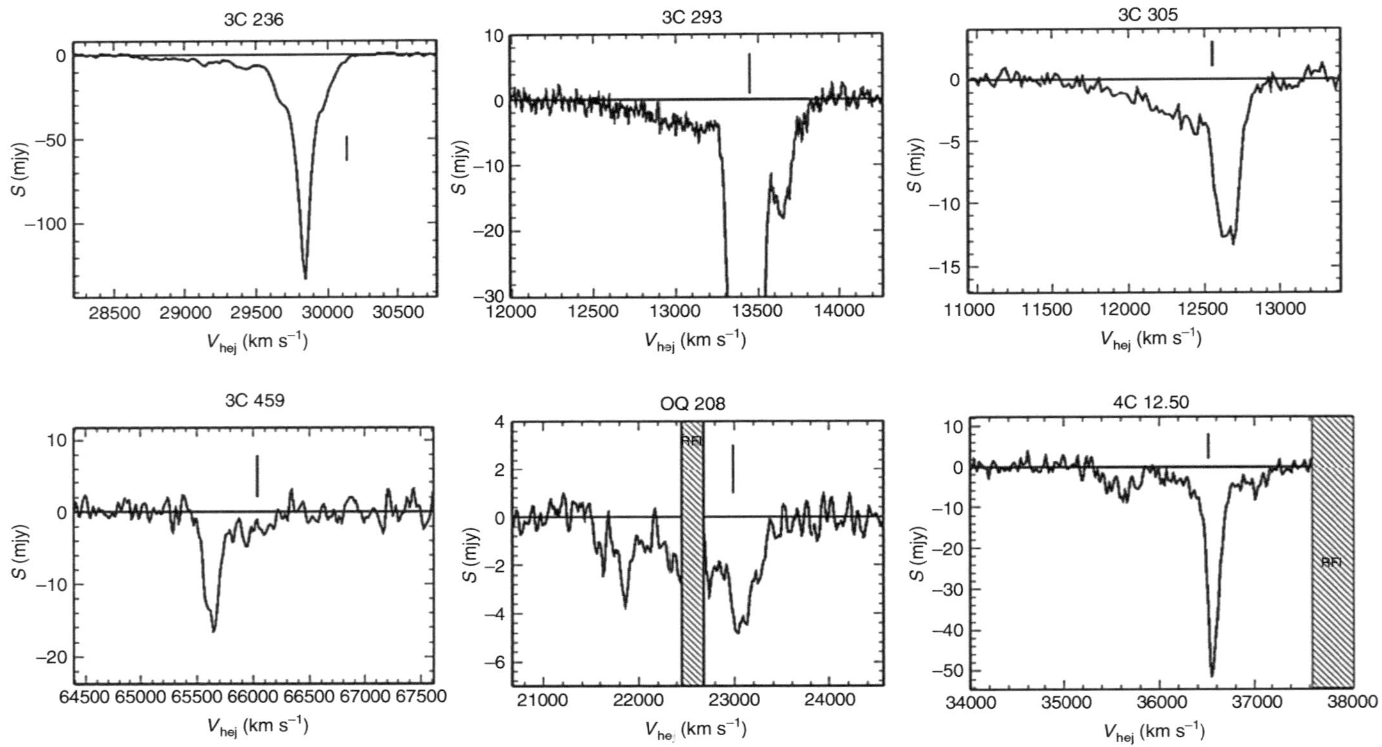

Fig. 4 Examples of broad HI absorption detected in radio galaxies, for more details see [20]

metals and inhibiting star formation in the AGN host galaxy. However, the HI 21cm absorption is extremely weak (peak optical depth typically $\sim 10^{-3}$) and wide ($\geq$1,000 km s^{-1}), and thus difficult to detect, requiring both high spectral dynamic range and wide bandwidths. Such absorption searches have hence only been possible against a few nearby, bright radio galaxies. It is important to determine whether such jet-driven HI outflows are a common phenomena in radio-loud AGN; if so, they could play an important role in recycling galactic material at all redshifts. Equally important is whether such outflows are detected at all phases of AGN evolution, or whether they are confined to the "radio mode", causing the suppression of cooling flows and the quenching of star formation in the host galaxy (e.g. [7]).

Finally, [22] and [26] have recently detected extremely broad HI 21cm absorption in the nearby radio source 4C37.11, which appears to be the kinematical signature of supermassive binary black holes (separated by only $\sim$7 pc) in the centre of the galaxy. This appears to be another interesting use of HI absorption to identify such extreme systems.

The detection obtained so far are limited to relatively nearby radio sources ($z < 0.9$, [27,31]). Interestingly, searches at higher redshift have been disappointing with so far only one detection at $z = 3.4$ (J0902 + 343, [4, 5], see Fig. 5). The reason for this lack of detections could be the limited sensitivity of the available of radio telescopes (at frequencies below $\sim$1,000 MHz) combined with the hostile RFI environment making this search not straightforward. However, it is important to point out that neutral hydrogen has been detected at high redshift via Lyα observations (e.g. [8, 33] and references therein). Furthermore, high-z galaxies are known to contain often molecular gas ([8, 16, 23, 24], and references therein). It is, therefore, important to understand whether the lack of HI 21cm detections indicate differences in the physical conditions of the gas in high-z radio sources.

3 Goals for the Next Generation HI Absorption Surveys and the Role of SKA

What do we need in order to make a major step forward in the study of HI 21cm absorption in radio sources and what should be the characteristics of the new radio telescopes that would allow this? The next generation of HI 21cm absorption surveys should aim at increasing the statistics by dramatically expanding the number of available HI absorption, looking at the neutral hydrogen content of the fainter population of radio sources (or sources with fainter radio cores) looking for differences in the structure of their central regions and in how these sources are fuelled, finding how common are fast outflows of neutral hydrogen, assessing their importance for feedback and galaxy evolution, extended the search of HI absorption to high-z and, finally, tracing the kinematics of the gas with high spatial resolution observations.

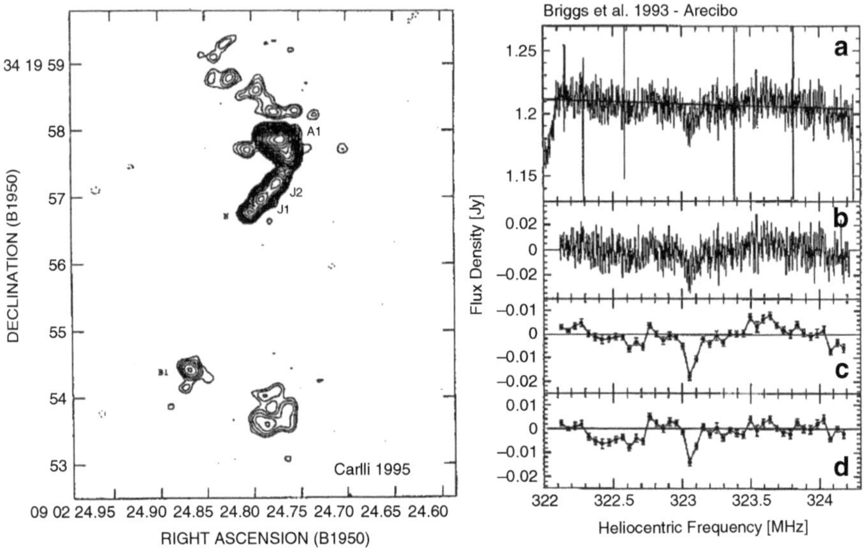

Fig. 5 Continuum image and HI absorption profile of the only source, J0902 + 343, detected at high redshift ($z = 3.4$)

A variety of "ingredients" are needed to reach these goals. The sensitivity is, of course, a key ingredient. By exploring low optical depth systems (also in weak radio sources) we aim at taking out many observational bias, e.g. investigate if there are difference in the gas content wrt to the different phases of evolution of the radio sources or for sources in different galactic environment. The area covered is also essential for a big step forward in building up large samples and to allow statistics of different type of radio sources at different redshifts. The spectral stability of the system, and in particular the bandpass stability, is an other important aspect (see below). The frequency coverage will be essential for investigating how the gas content changes wrt to redshift. Finally, high spatial resolution is also important for the study of HI 21cm absorption. Only by imaging the spatial distribution and kinematics of the neutral hydrogen it is possible to understand the origin of the absorber. Some of the goals listed above can be now explored with the surveys planned for the SKA pathfinders/precursors. For example, large area (or all sky) surveys are already planned or proposed with ASKAP and Apertif. Even the most shallow of theses surveys will provide many hundred detections of HI 21cm absorption with optical depth down to $\tau \sim 1\%$ for sources stronger than $\sim$100 mJy (column density $\sim 10^{20}\,\mathrm{cm}^{-2}$ for absorption of about 100 km/s). Further improvements, e.g. reaching lower optical depth, are expected to be achieved by using stacking techniques. The distribution of redshift for sources stronger than 100 mJy has been obtained from the simulations by [34] (S^3 – The SKA simulated sky, http://s-cubed.physics.ox.ac.uk/) for an area 400 deg^2 ($20° \times 20°$), see Fig. 6. While the SKA pathfinders/precursors will mainly allow to cover redshift range up

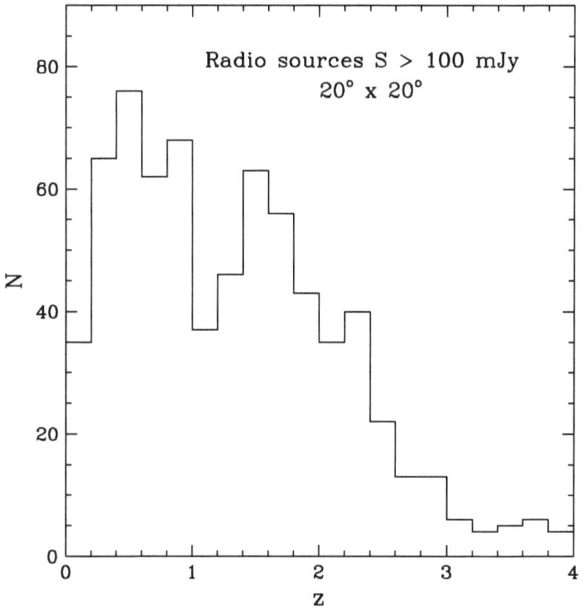

Fig. 6 Distribution of redshifts for radio sources stronger than 100 mJy in 400 deg^2 obtained from S^3 simulations [34]

to $z = 1$ (in the case of the ASKAP survey FLASH, PI E. Sadler), SKA and SKA Phase 1 will allow to expand this distribution to high redshifts.

The large surveys with the SKA pathfinders/precursors will already help improving our knowledge of the HI 21cm content and characteristics in AGN but a major step in sensitivity is also essential, and this is what SKA (hopefully already SKA Phase 1) should provide. Figure 7 shows the sensitivity to optical depth (τ) expected (at the 5σ limits) for various A_{eff}/T_{sys} with WSRT-Apertif and, ASKAP and EVLA as reference and SKA Phase 1 as planned is SKA memo 125 "*Concept Design for SKA Phase 1 (SKA$_1$)*" [12] also indicated.

The plot shows that in order to get a major improvement and being able to detect e.g. outflows in sources of a few hundred mJy flux, or 1% optical depth absorption in tens of mJy sources (much more common in deep surveys than bright sources) the proposed 1000 A_{eff}/T_{sys} is the minimum requirement.

In addition to the improvement in sensitivity, it is important to underlying the requirements in term of bandpass stability needed for HI 21cm absorption studies of strong radio sources. This is required in order to detect broad and shallow absorption features that we know that exist and they actually provide important information over the feedback process in AGN (tracing e.g. fast outflows), see Fig. 8 for an example [19]. The broad, shallow HI absorption associated with galactic outflow have very low optical depth ($\tau < 0.001$) and the detection require a bandpass stability to 10^4 or higher. Thus, the system should be stable (or only smoothly

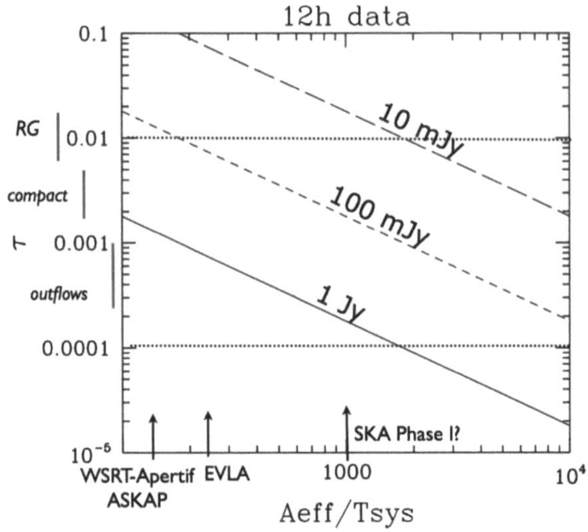

Fig. 7 Sensitivity to the optical depth of HI 21cm absorption as function of the sensitivity of various radiotelescopes. SKA Phase 1 is also marked

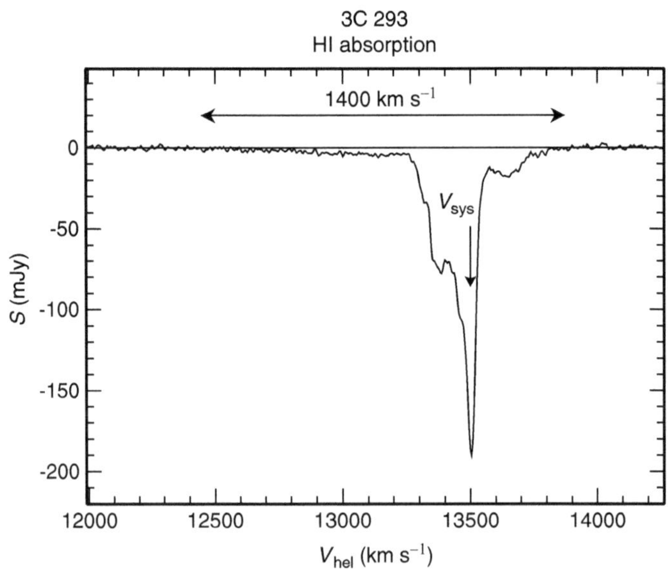

Fig. 8 Example of broad, shallow HI absorption as detected in 3C293 (see [19] for details). The broad component has been detected thanks to the broad band available and the stability of the band to a level of 10^4

changing!) with spectral dynamic range reaching at least 10^4, and even better in the case of SKA.

Finally, the study of HI absorption in AGN requires the broadest possible coverage to low frequencies so that the evolution of gas content as function of redshift can be traced. As stated in the *SKA Science Reference Mission*, we would like to be able to explore the presence of HI in radio sources up to redshift $z \sim 6$. Long baselines should also be considered in the planning for SKA in order to allow high enough spatial resolution to understand the details of the distribution and kinematics of the gas. Baselines of at least 3000 km (this is the number now reported in the *SKA Science Reference Mission*) will be needed in order to study the structure of the absorbing medium. These baselines would correspond to linear scales of ~10pc for nearby objects while for high redshift the resolution will be of ~200 pc for z ~ 2.

In summary, the study of associated HI absorption is already promising exciting results with the surveys planned for the SKA pathfinders/precursors and, even more, with the full SKA. However, the synergy with data from other facilities (large optical surveys, follow up using ALMA etc.) will be a crucial factor for a full success of these radio surveys.

References

1. Allen et al.: Mon. Not. R. Astron. Soc. **372**, 21 (2006)
2. Araya et al.: Astrophys. J. **139** 17 (2010)
3. Bellamy, M.J., Tadhunter, C.N.: Mon. Not. R. Astron. Soc. **353**, 105 (2004)
4. Briggs, F.H., Sorar, E., Taramopoulos, A.: Astrophys. J. **415**, L99 (1993)
5. Carilli, C.L.: Astron. Astrophys. **298**, 77 (1995)
6. Conway, J.E.: Int. Astron. Union Symp. **175**, 92 (1996)
7. Croton et al.: Mon. Not. R. Astron. Soc. **365**, 11 (2006)
8. De Breuck, C., et al.: Astron. Astrophys. **401**, 911 (2003)
9. Emonts, B.H.C. et al.: Mon. Not. R. Astron. Soc. **406**, 987 (2010)
10. Fanti, R. et al.: Astron. Astrophys. **231**, 333 (1990)
11. García-Burillo, S., Combes, F., Usero, A., Fuente, A.: Astron. Nachr. **330**, 245 (2009)
12. Garrett, M.A., Cordes, J.M., Deboer, D.R., Jonas, J.L., Rawlings, S., Schilizzi, R.T.: 2010, Proceedings of the ISKAF2010 Science Meeting. Published online at http://pos.sissa.it/cgi-bin/reader/conf.cgi?confid$=$112 (arXiv, arXiv:1008.2871)
13. Gupta, N., Salter, C.J., Saikia, D.J., Ghosh, T., Jeyakumar, S.: Mon. Not. R. Astron. Soc. **373**, 972 (2006)
14. Heckman et al.: Astron. J. **88**, 583 (1983)
15. Kanekar, N., Briggs, F.H.: New Astron. Rev. **48**, 1259 (2004)
16. Klamer, I.J., Ekers, R.D., Sadler, E.M., Weiss, A., Hunstead, R.W., De Breuck, C.: Astrophys. J. **621**, L1 (2005)
17. Liszt, H.S.: Astrophys. J. **371**, 865 (2001)
18. Morganti, R., Oosterloo, T.A., Tadhunter, C.N., van Moorsel, G., Killeen, N., Wills, K.A.: Mon. Not. R. Astron. Soc. **323**, 331 (2001)
19. Morganti, R., Oosterloo, T.A., Emonts, B.H.C., van der Hulst, J.M., Tadhunter, C.N.: Astrophys. J. **593**, L69 (2003)
20. Morganti et al.: Astron. Astrophys. **444**, L9 (2005)

21. Morganti, R., Oosterloo, T., Struve, C., Saripalli, L.: Astron. Astrophys. **485**, L5 (2008)
22. Morganti, R., Emonts, B., Oosterloo, T.: Astron. Astrophys. **496**, L9 (2009)
23. Nesvadba, N.P.H., et al.: Mon. Not. R. Astron. Soc. **395**, L16 (2009)
24. Peck et al: Astrophys. J. **521**, 103 (1999)
25. Roberts, M.: Astrophys. J. **161**, L9 (1970)
26. Rodriguez, C., Taylor, G.B., Zavala, R.T., Pihlström, Y.M., Peck, A.B.: Astrophys. J. **697**, 37 (2009)
27. Salter et al.: Astrophys. J. **715L**, 117 (2010)
28. Shostak et al.: Astron. Astrophys. **119**, L3 (1983)
29. Struve, C., Conway, J.: Astron. Astrophys. **513**, 10 (2010)
30. van Gorkom, et al.: Astron. J. **97**, 708 (1989)
31. Vermeulen, R. et al.: Astron. Astrophys. **404**, 861 (2003)
32. Veilleux et al.: Annu. Rev. Astron. Astrophys. **43**, 769 (2005)
33. Villar-Martín, M., et al.: Astron. Nachr. **327**, 187 (2006)
34. Wilman et al: Mon. Not. R. Astron. Soc. **388**, 1335 (2008)
35. Wolfe, A.M., Gawiser, E., Prochaska, J.X.: Annu. Rev. Astron. Astrophys. **43**, 861 (2005)

Galaxy Dynamics

W.J.G. de Blok, S.-H. Oh, and B.S. Frank

Abstract The Square Kilometre Array will revolutionize our understanding of the evolution of galaxies. In the very near future the various SKA precursors and pathfinders will also start surveying the sky, making their contributions to our knowledge. Recently large HI surveys on existing telescopes have already started to show us what will be possible in terms of relating gas densities to star formation, tracing cold accretion, and characterizing the dynamics of nearby galaxies. This contribution highlights some recent and ongoing work in characterizing the dynamics of the cold gaseous interstellar medium in nearby galaxies, as well as recent comparisons between de dynamics of real versus simulated dwarf galaxies.

1 Introduction

One of the Key Science Questions for the Square Kilometre Array (SKA) is *The Evolution of Galaxies – how do galaxies assemble and evolve?* The SKA will be able to directly trace the gradual transformation from primordial neutral hydrogen (HI) gas into galaxies over cosmic time. However, direct *detailed* observations of the sub-kpc-scale physical processes that cause this transformation, taking place both inside and around these evolving galaxies, will probably stay beyond our reach – even with the SKA – for a large span of cosmic time due to resolution and sensitivity problems. The only place where such a comprehensive survey of the "Galactic Ecosystem" can be made is the nearby universe; it is the only place where we can study, in detail, the flow of gas into galaxies, its physical conditions, its transformation into stars, and how it, in turn, is affected by feedback from these stars. The HI kinematics tell us about the distribution of dark matter, angular momentum, the shape of the halo potential, the disk gravitational potential, and,

W.J.G. de Blok (✉)
ACGC, Department of Astronomy, University of Cape Town, Rondebosch 7700, South Africa
e-mail: edeblok@ast.uct.ac.za

D. Barbosa et al. (eds.), *The Square Kilometre Array: Paving the way for the new 21st century radio astronomy paradigm*, Astrophysics and Space Science Proceedings, DOI 10.1007/978-3-642-22795-0_5, © Springer-Verlag Berlin Heidelberg 2012

ultimately, how the dark and visible matter together determine and regulate the evolution of galaxies. These local galaxies are the "fossil records" of the distant, high-redshift galaxies, and provide a wealth of information that can further refine models of galaxy formation and evolution. This knowledge can guide the interpretation of similar, but necessarily less detailed, observations at higher redshifts. The study of nearby galaxies thus provides the foundations on which studies of higher redshift galaxies must be built.

In this contribution we give a short summary of some ongoing work in characterizing the dynamics of the cold gaseous interstellar medium in nearby galaxies, as well as recent comparisons between de dynamics of real versus simulated dwarf galaxies.

2 Dark Matter in Real and Simulated Dwarf Galaxies

The "cusp/core" problem, the discrepancy between the predicted and observed dark matter distribution in the centres of galaxies has been intensively discussed from both observational and theoretical sides for almost two decades ever since the advent of high-resolution ΛCDM simulations. In particular, dwarf galaxies have been at the heart of the discussion since they are ideal objects for measuring the dark matter distribution near the centres of galaxies due to the small contribution of baryons to the total matter content (e.g., [20]). We have begun to address the "cusp/core" problem with 7 dwarf galaxies taken from THINGS (P.I.: F. Walter; [30]). Using novel techniques to extract the undisturbed kinematics of gas component, we derived mass models for the dark matter component of the galaxies and found that the mean value of the inner slopes of the mass density profiles is $\alpha = -0.29 \pm 0.07$ [17, 18]. As shown in the right panel of Fig. 1, these rather flat slopes show a significant discrepancy with the steep slope of ~ -1.0 predicted from previous dark-matter-only simulations. The exquisite data used in this study allowed unprecedented treatment of the effects of observational uncertainties, such as beam smearing, center offset and non-circular motions, which may play a role in hiding central cusps [1, 2, 6, 17, 22, 24–26, 28]. The results have thus significantly strengthened the observational evidence that the dark matter distribution near the centers of dwarf galaxies follows a near–constant density core.

Before using these results as a repudiation of cold dark matter, however, one must also examine carefully the modeling on which the central cusp predictions are based, which are usually done using N-body simulations that include only the effects of gravity on structure formation. Although baryons make up only $\sim$14% of the matter of the Universe, this dissipative constituent of the Universe cools with cosmological time and accumulates in the central regions of DM halos, making up a dynamically important fraction. Several mechanisms have been proposed whereby these central baryons can affect the central cusp–like dark matter distribution which is found in pure dark matter simulations. A rapid change in potential (faster than the dynamical time) due to star burst triggered outflows, is one mechanism which

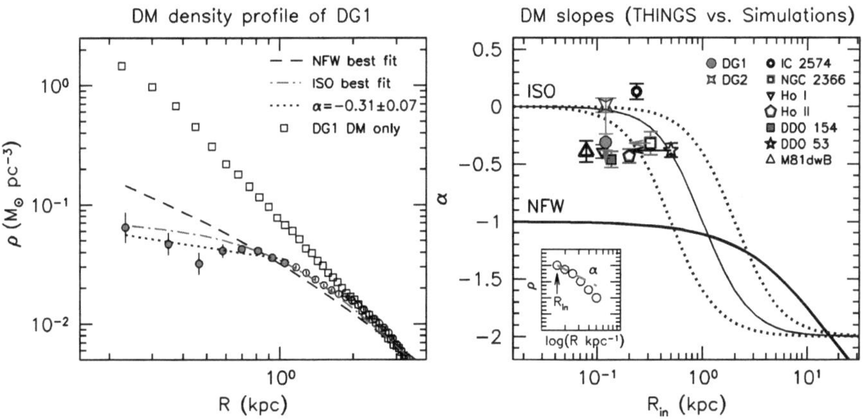

Fig. 1 *Left*: The dark matter density profile of the simulated dwarf galaxy, DG1. The *circles* represent the dark matter density profile derived using the same analysis techniques and tools as done in the THINGS dwarf galaxies [18]. The *open rectangles* indicate the profile from the dark-matter-only simulation. The *dashed* and *dash-dotted lines* indicate the best fitted NFW and pseudo-isothermal (ISO) halo models, respectively. The inner slope of the profile is measured by a least squares fit (*dotted lines*) to the data points less than 1.0 kpc as indicated by *gray dots*. The measured inner slope α is shown in the panel. *Right*: The inner slope of the dark matter density profile plotted against the radius of the innermost point. The inner density slope α is measured by a least squares fit to the inner data point as described in the small figure. The inner slopes of the mass density profiles of the 7 THINGS dwarf galaxies significantly deviate from the $\alpha \sim -1.0$ predicted from dark-matter-only simulations. Instead, they are better consistent with the simulated dwarf galaxies (DG1 and DG2) from a set of SPH+N–body simulations by [8] where the baryonic feedback processes are included [19]

has been shown to be capable of transforming cusp–like profiles into (flatter) cores ([16, 21] although see [7]). Supernova driven random bulk motions of gas in proto-galaxies [12] has also been shown in models to flatten cusps, as have the effects of dynamical friction acting on gas clumps [5], and the transfer of angular momentum from baryons to the dark matter [27]. Modeling an inhomogeneous multi-phase, interstellar medium is critical for simulating the baryonic feedback processes in galaxies [3, 23]. Yet cosmological simulations have not, until recently, been able to achieve enough resolution to model even such an inhomogeneous ISM, but have been forced to treat the important processes of star formation and feedback as "sub-grid" physics, averaging star formation and supernova feedback over large volumes, compared to the typical structural scales (<1 kpc) of small galaxy disks.

Most recently, [8] have performed high-resolution cosmological N–body + Smoothed Particle Hydrodynamic (SPH) simulations of dwarf galaxies under the ΛCDM paradigm, that include the effect of baryonic feedback processes, such as gas cooling, star formation, cosmic *UV* background heating and most importantly physically motivated gas outflows driven by SNe. The major finding of [8] was that once star formation is associated with high density gas regions, a significant amount of baryons with low angular momentum is efficiently removed by strong

SNe–driven injection of thermal energy and the following gas outflows that carry an amount of gas at a rate of two to six times the local star formation rate. The large scale outflows in turn induce two effects: the loss of low angular momentum gas from the central regions prevents the formation of bulges in low mass systems. Secondly, the clumpy nature of the gas and rapid ejection on short timescales has dynamical effects on the dark matter potential, creating a shallower density profile (see [3, 12, 13, 15]) Moreover, the simulated galaxies have a $z = 0$ baryonic budget consistent with photometric and kinematic estimates [14, 29]. The kinematic properties of the simulated dwarf galaxies (DG1 and DG2) are very similar to those of the THINGS dwarf galaxies in terms of their maximum rotation velocity ($\sim$60 km s^{-1}) and dynamical mass ($\sim 10^9 M_{\odot}$), allowing a direct comparison between the simulations and observations to be made.

In order to examine how the baryonic feedback processes affect the dark matter distribution at the centers of dwarf galaxies, we have analyzed the simulated galaxies in a homogeneous and consistent manner as described in [18]. The techniques used in deriving dark matter density profiles were found to provide accurate results when compared with the true underlying profiles, supporting the veracity of the techniques employed by observers. Therefore, this provides a quantitative comparison between the simulations and the observations.

From this, we test the general predictions from ΛCDM simulations: (1) the steep rotation curve inherent in the central cusp, and (2) the steep inner slope of ~ -1.0 of the dark matter density profiles. We find that the dark matter rotation curves of the newly simulated dwarf galaxies rise less steeply at the centers than those from dark-matter-only simulations. Instead, they are more consistent with those of the THINGS dwarf galaxies. In addition, the mean value of the inner density slopes α of the simulated dwarf galaxies is $\simeq -0.4 \pm 0.1$. Compared to the steep slope of ~ -1.0 predicted from the previous dark-matter-only simulations (including our simulations run with DM only), these flat slopes are in better agreement with $\alpha = -0.29 \pm 0.07$ found in the 7 THINGS dwarf galaxies analysed by [18].

These confirm that energy transfer and subsequent gas removal in a clumpy ISM have the net effect of causing the central DM distribution to expand, while at the same time limiting the amount of baryons at the galaxy center. By the present time the DM central profile in the simulated dwarf galaxies is well approximated by a power law with slope α of $\sim -0.4 \pm 0.1$. These values of α are significantly flatter than in the collisionless control run and are in agreement with those of observed shallow DM profiles in nearby dwarf galaxies.

3 Comparisons of CO and HI Dynamics in THINGS Galaxies

3.1 Introduction

Large HI surveys have mapped the neutral gas in local galaxies to low column densities with high spectral and spatial resolution. As such, studies of HI have been used to extract a variety of properties, such as star-formation rates, gas dynamics and, hence, the dark matter content in gas-rich galaxies. Molecular gas is the missing link between the neutral gas and stars, and the content and dynamics of molecular gas in nearby galaxies have been inferred through observations of the CO transitions, which is the most accesible tracer of H_2. The majority of these observations, however, have lacked the sensitivity and area-coverage attainable with HI observations, prohibiting a single, consolidated study of the molecular gas dynamics of nearby galaxies and comparisons with HI.

Recent advances in high frequency radio receiver design has lead to more sensitive observations of CO in nearby galaxies, allowing for a mapping of the molecular gas content along the extent of the HI disk.

The aim of the current work is to provide a comparison between the gas dynamics of HI and CO in nearby galaxies, as observed with recent, high resolution surveys. The sample galaxies are part of the The HI Nearby Galaxy Survey (THINGS, [30]). In THINGS observations, the spectral resolution is $5\,\mathrm{km\,s^{-1}}$, and the spatial resolution $10''$. 34 gas-rich galaxies of various morphologies were observed. Of these 34 galaxies, a dynamical analysis was performed on 19 ([4]).

The HERA CO Line Extragalactic Survey (HERACLES, [11]) is a survey of the molecular gas content of these galaxies, as measured by observing the CO $(J = 2 \rightarrow 1)$ line. The spectral and spatial resolution of the HERACLES observations are $2.6\,\mathrm{km\,s^{-1}}$ and $13''$ respectively and is comparable to that of the THINGS survey, allowing for a comparison of the dynamics of both the neutral and molecular gas components.

Of the 34 galaxies observed in THINGS, 27 galaxies were observed in HERACLES. Of these, there are 15 galaxies which were analysed in the HI by [4], of which 12 were detected in CO. These are the galaxies on which we focus our present study.

We also briefly explore the interchangeability of CO and HI in the Tully–Fisher relation (TFR), as suggested in [10].

3.2 *Methods and Techniques*

We derive a peak velocity-field by fitting Gauss–Hermite polynomials of order 3 to each pixel in a CO data cube, in accordance with the method prescribed in [4].

We also derive the difference velocity-field by subtracting the CO velocity-field from the HI velocity-field derived in [4].

We compute two rotation curves from the CO velocity-field using a tilted ring model with the `GIPSY` program `ROTCUR` for each galaxy. The first is computed by simply applying the geometrical parameters as derived from the THINGS data (the THINGS-Model) and as presented in [4]; the second is computed using an independent `ROTCUR` analysis where we solve for all the dynamical parameters, in addition to the circular velocity (the All-Free case). We assume only circular rotation.

3.3 Results for NGC 925, 2403 and 5055

We present a summary of results from a representative subsample of galaxies (NGC 2403, NGC 5055 and NGC 925).

NGC 2403 is a late-type gas rich spiral. The velocity and difference velocity-fields are plotted in Fig. 2. The difference velocity-field indicate that the differences between the CO and HI are small and localized. This is to be expected, since the CO emission is likely to be associated with regions of turbulence and star-formation. Figure 3 shows that the CO THINGS-Model rotation curve agrees quite closely with that of the HI, out to a radius of about 150 arcsec. The CO All-Free rotation curve shows a similar trend, and the departure between the HI and CO rotation curves correlates with a low filling-factor of the tilted rings.

NGC 5055 is a late type LINER with flocculent spiral arms evident in the HI and in the CO. The velocity and difference-velocity-fields are plotted in Fig. 4. The difference-velocity-field shows that there is a good agreement between the HI and the CO velocities. Part of the differences in the inner parts of the galaxy can be attributed to a rapidly rotating molecular circum-nuclear disk.

There is a good agreement between the HI and the CO-All-Free rotation curves from 30 arcsec outwards to the extent of the CO emission.

NGC 925 is a late-type barred galaxy. Figure 5 shows the velocity-fields of the CO and HI respectively. A comparison with the SINGS [9] image and Fig. 5 shows that the CO emission is limited to a region along the bar on the approaching side of the galaxy presenting an additional challenge in rotation curve derivation. The low CO intensity and the concentration of CO along the bar means that it is not straightforward to derive a rotation curve that is an accurate represention of the rotational velocity.

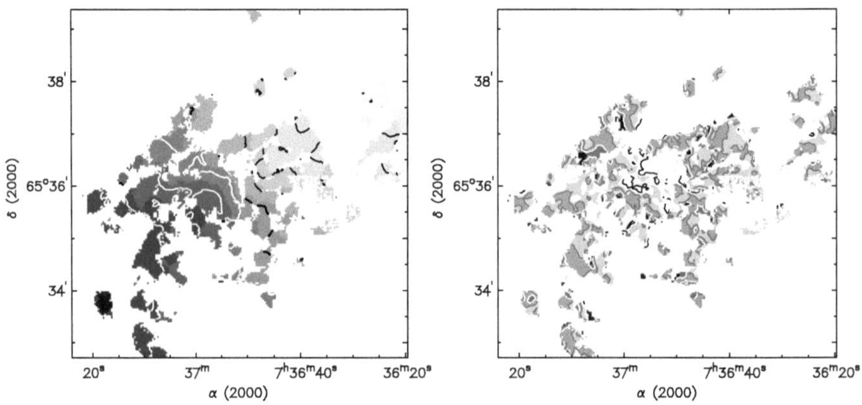

Fig. 2 NGC 2403 CO (*left panel*) and HI-CO (*right panel*) difference velocity-fields. In the CO velocity-field, levels are from 0 km s^{-1} to 225 km s^{-1} in steps of 25 km s^{-1}, with v_{sys} = 132.8 km s^{-1}. Difference velocity-field: Grayscale is plotted from −10 km s^{-1} to 10 km s^{-1} in steps of 5 km s^{-1}. *Black* contours correspond to an absolute difference of 10 km s^{-1}, *white* to 5 km s^{-1} and *dark grey* to 0 km s^{-1}

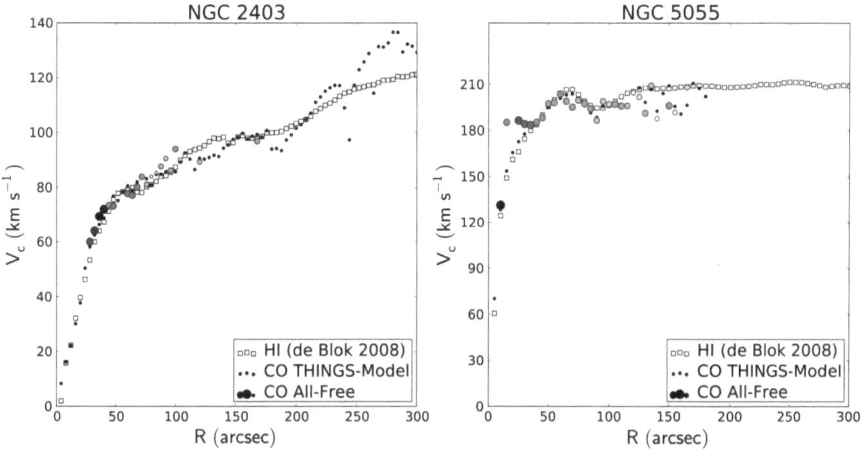

Fig. 3 NGC 2403 and NGC 5055 CO rotation curves. The size and color of the CO All-Free points scale with the filling factor. Here we plot points with a filling factor greater than 5%

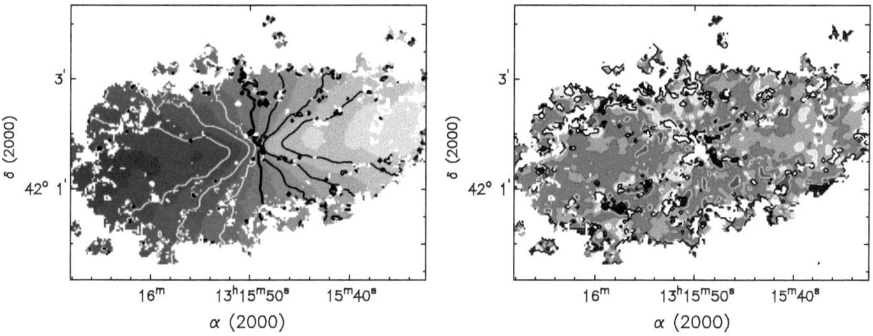

Fig. 4 NGC 5055 CO velocity-fields, as in Fig. 2. In the CO velocity-field, levels are from 225 km s^{-1} to 725 km s^{-1} in steps of 50 km s^{-1}, with v_{sys} = 496.8 km s^{-1}. Contours in the difference field are as in Fig. 2

3.4 Summary

There is a good agreement between the dynamics of the HI and the CO for the galaxies presented. The THINGS-Model parameters are a suitable first estimate for the CO dynamics, but independent rotational parameters should be derived from the CO data. Local differences can be attributed to regions of turbulence in the ISM, as well as the additional motions due to a possible circum-nuclear disk in NGC 5055 and the bar in NGC 925.

The HI and CO global profiles are plotted in Fig. 6. We see that substantial differences in the velocity widths between both components are possible, which implies that a CO TFR needs to be calibrated independently of an HI TFR. This is

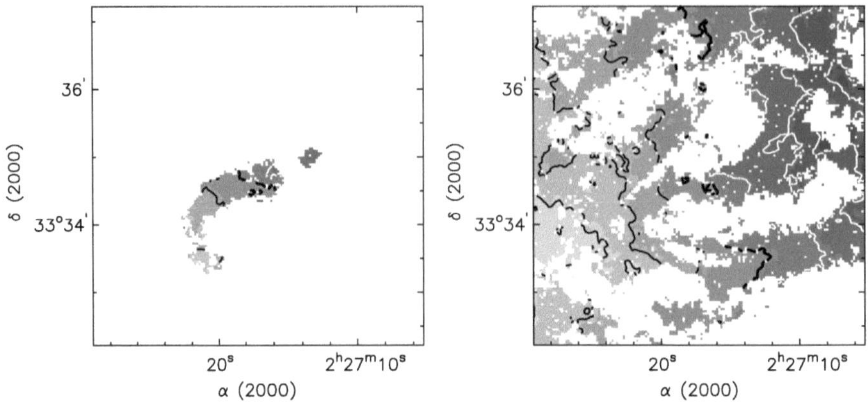

Fig. 5 NGC 925 CO (*left panel*) and HI (*right panel*) velocity-fields. Levels are from 425 km s^{-1} to 700 km s^{-1} in steps of 25 km s^{-1}, with $v_{sys} = 546.3$ km s^{-1}

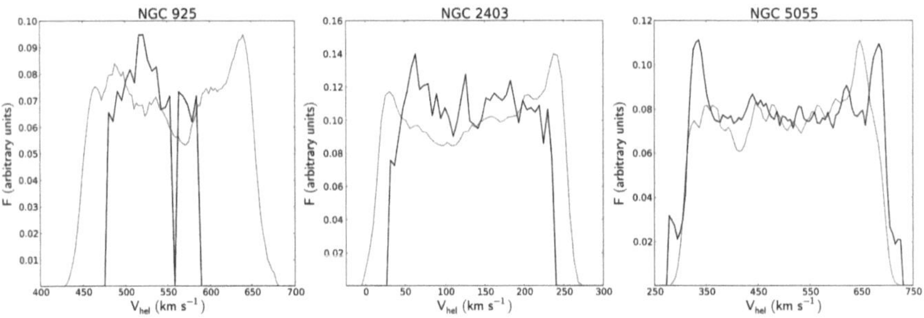

Fig. 6 HI (*grey*) and CO (*black*) global profiles for NGC 925, NGC 2403 and NGC 5055

important when considering the observations of high-redshift CO as a proxy for HI in a general TFR.

In future work we will present rotation curves of all the galaxies analyzed in this study (Frank et al., in preparation), where we compare the dynamics of the CO and the HI, in addition to constructing new mass-models including the molecular gas. This work aims to provide a template for future studies of HI and CO with next-generation instruments (e.g. the SKA and ALMA), since we can associate differences in the morphologies of global profiles of HI in CO in nearby galaxies with an in-depth comparison of the dynamics.

References

1. Blais-Ouellette, S., Carignan, C., Amram, P., Côté, S.: Astron. J. **118**, 2123 (1999) DOI 10.1086/301066
2. Bolatto, A.D., Simon, J.D., Leroy, A., Blitz, L.: Astrophys. J. **565**, 238 (2002)

3. Ceverino, D., Klypin, A.: Astrophys. J. **695**, 292 (2009)
4. de Blok, W.J.G., Walter, F., Brinks, E., Trachternach, C., Oh, S.-H., Kennicutt, R.C.: Astron. J. **136**, 2648 (2008)
5. El-Zant, A., Shlosman, I., Hoffman, Y.: Astrophys. J. **560**, 636 (2002)
6. Gentile, G., Burkert, A., Salucci, P., Klein, U., Walter, F.: Astron. J. **634**, L145 (2005)
7. Gnedin, O.Y., Zhao, H.: Mon. Not. R. Astron. Soc. **333**, 299 (2002)
8. Governato, F., Brook, C., Mayer, L., Brooks, A., Rhee, G., Wadsley, J., Jonsson, P., Willman, B., Stinson, G., Quinn, T., Madau, P.: Nature **463**, 203 (2010)
9. Kennicutt, R.C., Jr., et al.: Publ. Astron. Soc. Pac. **115**, 928 (2003)
10. Lavezzi, T.E., Dickey, J.M.: Astron. J. **116**, 2672 (1998)
11. Leroy, A.K., et al.: Astron. J. **137**, 4670 (2009)
12. Mashchenko, S., Couchman, H.M.P., Wadsley, J.: Nature **442**, 539 (2006)
13. Mashchenko, S., Wadsley, J., Couchman, H.M.P.: Science **319**, 174 (2008)
14. McGaugh, S.S., Schombert, J.M., de Blok, W.J.G., Zagursky, M.J.: Astrophys. J. **708**, 14 (2010)
15. Mo, H.J., Mao, S.: Mon. Not. R. Astron. Soc. **353**, 829 (2004)
16. Navarro, J.F., Frenk, C.S., White, S.D.M.: Astrophys. J. **462**, 563 (1996)
17. Oh, S.H., de Blok, W.J.G., Walter, F., Brinks, E., Kennicutt, R.C.: Astron. J. **136**, 2761 (2008)
18. Oh, S., de Blok, W.J.G., Brinks, E., Walter, F., Kennicutt, R.C., Jr.: ArXiv e-prints (2010)
19. Oh, S., Brook, C., Governato, F., Brinks, E., Mayer, L., de Blok, W.J.G., Brooks, A., Walter, F.: ArXiv e-prints (2010)
20. Prada, F., Burkert, A.: Astrophys. J. Lett. **564**, 73 (2002)
21. Read, J.I., Gilmore, G.: Mon. Not. R. Astron. Soc. **356**, 107 (2005)
22. Rhee, G., Valenzuela, O., Klypin, A., Holtzman, J., Moorthy, B.: Astrophys. J. **617**, 1059 (2004)
23. Robertson, B.E., Kravtsov, A.V.: Astrophys. J. **680**, 1083 (2008)
24. Simon, J.D., Bolatto, A.D., Leroy, A., Blitz, L.: Astrophys. J. **596**, 957 (2003)
25. Spekkens, K., Giovanelli, R., Haynes, M.P.: Astron. J. **129**, 2119 (2005) DOI 10.1086/429592
26. Swaters, R.A., Madore, B.F., van den Bosch, F.C., Balcells, M.: Astrophys. J. **583**, 732 (2003)
27. Tonini, C., Lapi, A., Salucci, P.: Astrophys. J. **649**, 591 (2006)
28. van den Bosch, F.C., Robertson, B.E., Dalcanton, J.J., de Blok, W.J.G.: Astron. J. **119**, 1579 (2000)
29. van den Bosch, F.C., Burkert, A., Swaters, R.A.: Mon. Not. R. Astron. Soc. **326**, 1205 (2001)
30. Walter, F., Brinks, E., de Blok, W.J.G., Bigiel, F., Kennicutt, R.C., Thornley, M., Leroy, A.: Astron. J. **136**, 2563 (2008)

Transient Phenomena: Opportunities for New Discoveries

T. Joseph W. Lazio

Abstract Known classes of radio wavelength transients range from the nearby (stellar flares and radio pulsars) to the distant Universe (γ-ray burst afterglows). Hypothesized classes of radio transients include analogs of known objects, such as extrasolar planets emitting Jovian-like radio bursts and giant-pulse emitting pulsars in other galaxies, to the exotic, such as prompt emission from γ-ray bursts, evaporating black holes and transmitters from other civilizations. Time domain astronomy has been recognized internationally as a means of addressing key scientific questions in astronomy and physics, and pathfinders and Precursors to the Square Kilometre Array (SKA) are beginning to offer a combination of wider fields of view and more wavelength agility than has been possible in the past. These improvements will continue when the SKA itself becomes operational. I illustrate the range of transient phenomena and discuss how the detection and study of radio transients will improve immensely.

1 Introduction

The field of radio astronomy, and more broadly the recognition that multi-wavelength astronomy is important, resulted from a study of transient radio emissions. In the early part of the twentieth century, Bell Labs was interested in determining the source of transient radio interference on trans-atlantic radio communication links. This interest in transient radio interference resulted in a

T.J.W. Lazio (✉)
Jet Propulsion Laboratory, California Institute of Technology, Mail Code 314-308, 4800 Oak Grove Dr., Pasadena, CA 91109, USA

Square Kilometre Array Program Development Office, University of Manchester, Manchester, UK
e-mail: Joseph.Lazio@jpl.nasa.gov

D. Barbosa et al. (eds.), *The Square Kilometre Array: Paving the way for the new 21st century radio astronomy paradigm*, Astrophysics and Space Science Proceedings, DOI 10.1007/978-3-642-22795-0_6,

young engineer, Karl Jansky, building a radio receiver from which he was able to isolate various sources of interference. Two of these sources were indeed transients, related to radio emission from lightening, but the other source of "interference" was a steady source of celestial radio waves toward the center of the Milky Way Galaxy [30].

While radio transients motivated Jansky's work, the celestial radio emission that he discovered was itself steady. Indeed, a New York Times headline announcing the discovery noted that its constancy was likely an indication that the emission was not the result of an extraterrestrial civilization. Subsequently, much of the work in radio astronomy assumed, either implicitly or explicitly, that the radio sky was largely unchanging. This assumption was consistent with the history of astronomy in which, with the exception of rare and dramatic events such as novae or supernovae, the sky was static.

A strong hint that the radio sky might be more dynamic than initially assumed was the discovery of radio pulsars [27], which was awarded the Nobel Prize. With observations now covering essentially the entire electromagnetic spectrum, there has also been a growing recognition that the time domain represents a largely unexplored part of natural parameter space in astronomy. This recognition has been re-inforced by a series of observations over the past decade that have revealed a host of suprises regarding known classes of sources as well as potentially the discovery of new classes of sources. The recent European ASTRONET and US Astronomy and Astrophysics Decadal Survey have emphasized the potential of time domain astronomy. The European ASTRONNET process produced a science vision for European astronomy [20], structured as a set of questions, for which time domain observations play a key role in answering. The US Decadal Survey, *New Worlds, New Horizons in Astronomy and Astrophysics* [5], explicitly named time-domain astronomy as a "science frontier discovery area."

The Key Science Program of the Square Kilometre Array has long encouraged a design philosophy for the telescope such that discovery of new sources and new phenomena is enabled [11], and the dynamic radio sky may be one avenue for such discoveries [15]. More broadly, that the SKA will study transient radio emitters is in keeping with a trend in astronomy, across wavelength, for telescopes to search for transient sources, both electromagnetically (e.g., *Swift*, SkyMapper, LSST) and gravitationally (LIGO, VIRGO). Using examples and discoveries from the past decade, I illustrate how transients at radio wavelengths that might be studied with the SKA address some of the larger astronomical questions raised by both ASTRONET and *New Worlds, New Horizons in Astronomy and Astrophysics*. This selection is necessarily incomplete, but I hope that illustrates the breadth of possible topics that time-domain radio observations can address.

2 Lighthouse-Like Brown Dwarfs

There is a strong correlation between the radio and X-ray luminosities of solar flares and stellar flares from main sequence stars, the so-called Güdel-Benz relation [21]. This relation holds across a range of stellar spectral classes, but it is becoming increasingly clear that very low mass stars and sub-stellar objects deviate, by potentially large factors [10^3 or more, 3].

Dramatic indications of the extent to which brown dwarfs violate this relation have been intense radio flares observed from various objects, including pulsed radio emission from some brown dwarfs not unlike the emission from pulsars [2, 4, 22]. This pulsed emission is consistent with an electron cyclotron maser operating in the magnetic polar regions of these objects [23]. This emission process is the same one that produces the intense radio emission from the giant planets and Earth in the solar system, suggesting that extrasolar planets might also be detectable as transient radio sources [33].

These observations indicate that there are significant changes in the structure of the coronae, magnetic fields, or both of stars as one considers stars of later and later spectral classes. In addition to being of interest in its own right, such astrophysical studies may also contribute to a better understanding of the Sun's corona and magnetic field ("What drives Solar variability on all scales?" Astronet). These play an important role in controlling the space environment around the Earth, upon which our technological civilization is increasingly dependent (e.g., satellite communications, navigation).

From the SKA perspective, many of the existing observations of the transient radio emission from very low mass stars and brown dwarfs has been at the "mid frequencies," typically between 5 and 10 GHz. An unexplored regime is at lower frequencies, below 1 GHz, and the analysis of recent observations of the planet HD 80606b suggest that observations below 100 MHz will be necessary to detect extrasolar planets [34]. Much of the existing work has also targeted known brown dwarfs [or extrasolar planets, but see 35], yet there may also be merit in all-sky surveys in an effort to find faint brown dwarfs that have escaped detection in previous infrared surveys.

3 Rotating Radio Transients (RRATs)

The ideal pulsar is a magnetized neutron star with a stable rotation that produces a completely regular pulse train. Soon after their discovery, however, it became clear that few, if any, pulsars obtained this ideal. Individual pulses vary in amplitude, both intrinsically as well as from propagation effects [44]; some pulsars emit "giant pulses" (§4), and some pulsars exhibit "glitches" in which the pulsar rotation period actually changes [7, 47].

Taken to an extreme, one might ask if there are "pulsars" that emit only single, or very few, radio pulses. A re-analysis of a pulsar survey, in which the search was optimized for finding single pulses, did indeed find such pulses, from a class of objects now known as rotating radio transients [RRATs, 38]. The identification of only a few objects, coupled with the realization that many more such objects had likely been overlooked in previous surveys, indicated that the Galactic population of neutron stars likely had been underestimated by a significant factor (2 or more), possibly resulting in tension with the estimated Galactic supernova rate [32]. Coupled with the increasing number of other classes of neutron stars (e.g., magnetars, accreting X-ray pulsars, compact central objects), the discovery of RRATs ties into larger questions of the possible end states of stellar collapse and the conditions during supernovae or in the progenitor stars that give rise to the diversity of neutron stars ("How do supernovae and gamma-ray bursts work?" Astronet; "What controls the masses, spins, and radii of compact stellar remnants?" NWNH).

From an SKA perspective, searches for pulsars, emitting either regular pulse trains or single pulses, are an important part of the SKA Key Science Program. These searches are likely to focus on frequencies around 1–2 GHz, in an effort to balance between lower frequencies (which optimizes for the typically steep spectra of pulsars) and higher frequencies (which mitigates interstellar propagation effects).

4 Probing the Intergalactic Medium

The majority of baryons in the current epoch are thought to be in large-scale ionized filaments that form "strands" of a "cosmic web" [12, 18, 19]. Observations of highly ionized species of oxygen and neon by both the *FUSE* and *Chandra* observatories are considered to be validations of these predictions. Absorption observations along various lines of sight suggest a diffuse medium with a temperature of order 10^6 K with a density of order $5 \times 10^{-5}\,\mathrm{cm}^{-3}$ [e.g., 28, 46]. While striking, these observations still suffer from the difficulty of probing only trace elements. The ionized hydrogen in the WHIM has not been detected directly.

The Crab pulsar emits so-called "giant" pulses – pulses with strengths 100 or even 1000 times the mean pulse intensity, at times out-shining the Crab Nebula itself [24]. For many years, this phenomenon was thought to be uniquely characteristic of the Crab pulsar, but giant pulses have since been detected from the millisecond pulsars PSR B1937+21 and PSR B1821−24 [14,45] and PSR B0540−69, the Crab-like pulsar in the Large Magellanic Cloud [31].

Cordes et al. [16] and McLaughlin et al. [39] assess how far away giant-pulse emitting pulsars could be detectable. They find that, even with existing instrumentation (e.g., Arecibo, GBT), giant-pulse emitting pulsars may be detectable over intergalactic distances ($\sim$1 Mpc). Because a plasma is a dispersive medium, any broadband radio pulses directly encode the electron (or plasma) column density in their frequency behavior. Consequently, detecting pulses from pulsars in other galaxies provides a direct measure of the intergalactic (ionized) hydrogen column density, and potentially of the local "cosmic web" ("Where are most of the metals throughout cosmic time?" Astronet).

From the SKA perspective, searches potentially can be conducted to much larger distances, encompassing more galaxies. Such searches will likely be conducted between about 0.4 and 1 GHz in order to balance the typically steep spectra of pulsars with the expected dispersion. Contrary to most pulsar searches, for which one objective is often to minimize the effects of dispersion, for studying the intergalactic medium, finding dispersion effects would be an advantage.

5 Electromagnetic Counterparts to Gravitational Wave Sources

There are a handful of double neutron star binary systems known. In at least two notable cases—PSR B1913+16, the "Hulse-Taylor" binary, and the more recently discovered "double pulsar" PSR J0737−3039—the orbits are decaying by an amount that is consistent with the emission of gravitational radiation from the systems [9, 50]. Estimates are that these systems will merge within the next roughly 0.1–1 Gyr, depending upon the system, with the expectation that the merger will generate gravitational waves [1]. While no gravitational waves from merging neutron star systems have been detected yet, it is expected that these systems represent a significant source population for ground-based laser interferometric instruments.

There are expectations that a merger of a double neutron star system could produce not only a gravitational wave signal but an electromagnetic one as well. Hansen and Lyutikov [25] make specific predictions based on certain assumptions about such a binary, but one might more generally expect that many gravitational wave sources could also produce electromagnetic wave signals [6]. Detecting objects both electromagnetically and gravitationally would increase dramatically the variety of information that could be extracted, such as greatly improved luminosity measurements (or limits) or detailed constraints on the physics of the gravitational wave event ("Can we observe strong gravity in action?" Astronet; "Discovery: Gravitational wave astronomy," NWNH).

From the SKA perspective, it is likely to be operational during a time when there will be both ground- and space-based gravitational wave observatories operating. Depending upon the success of Precursors and pathfinders, there are at least two operational models for the SKA to conduct coordinated gravitational-electromagnetic wave observations. In one mode, the SKA could follow-up gravitational wave alerts, either studying the sources themselves if their positions are well known or by conducting small "surveys" to search for transients or variable sources within the uncertainty region of the gravitational wave event. Alternately, the SKA could provide the source position, and other information, on potential gravitational wave events (e.g., particular classes of transients) that would enable gravitational wave detection to proceed with higher confidence.

6 Exotica

The sources described above belong generally either to populations of sources not originally throught to display transient behavior at all or are relatively straightforward extensions of known populations. There have also been discussions in the literature of what might be termed exotic objects. As examples, I highlight two classes of sources here.

Building on predictions of black hole evaporation [26], [43] noted that the particles produced during the evaporation of a small black hole could interact with an ambient magnetic field (interstellar or intergalactic) and produce a radio pulse. Within some reasonable assumptions about the ambient magnetic field and the particle densities, such radio pulses could be detected from essentially anywhere in the Galaxy. Interestingly, the typical radio wavelength for the pulse scales as the magnetic field strength as $\lambda \sim 10\,\mathrm{cm}(B/5\,\mu\mathrm{G})^{-2/3}$. Given that the intergalactic magnetic field is lower than the interstellar field, searching at longer wavelengths would tend to favor finding black holes in intergalactic space. No such pulses have yet been found [41,42]. One of the challenges of identifying such pulses could be that, by their nature, they would not be reproducible [10,37]. Positive identification might rely on obtaining a large sample of pulses, from which other characteristics could be determined.

Soon after the widespread adoption of radio as a communication medium, it was recognized that it might be possible to use radio transmissions to communicate not only between countries but also between worlds [13]. Since that time, there have been numerous searches for radio transmissions [48]. While none have been successful to date, the radio band of the spectrum still represents a potentially fruitful exploration area: Among other considerations, only at radio and gamma-rays is the Milky Way Galaxy transparent. Because of limited sensitivity, previous searches for radio transmissions would have been able to detect only "beacons," signals intentionally aimed in our direction, and sufficiently distant beacons might be transient due to propagation effects [17]. However, with the sensitivity of the SKA, signals with strengths comparable to our strongest radars become detectable over interstellar distance, so-called "leakage" signals. Such signals might very well be transient because, for instance, the transmitter might be fixed to a rotating planet. As the beam of the transmitter swept through the sky, the SKA would be able to detect it only when illuminated (analogous to a pulsar). Detection of an extraterrestrial civilization would help address one of the most enduring questions of humanity.

7 Unknown Classes of Sources

The past decade has also seen the discovery of sources that have not yet been identified, of which there are two notable examples. The first example was GCRT J1745−3009, a transient discovered at 0.3 GHz in observations toward the Galactic center [29]. This source exhibited a series of outbursts, approximately 10 min in duration with a periodicity of 77 min. No counterpart was detected at other wavelengths, and speculations for the explanation of this source have included a brown dwarf or very low mass star, a so-called white dwarf pulsar, and a double neutron star binary [29,49,51].

The second was a set of 10 transients discovered in nearly 1,000 epochs over 22 year of observations at 5 and 8.4 GHz [8]. There is insufficient evidence to classify these transients, specifically no multi-wavelength counterparts, and they need not all belong to the same class of object. Possible explanations for these transients include orphan gamma-ray burst afterglows, isolated neutron stars, stellar sources, and extreme scattering events [8, 40].

8 SKA Science Pathfinding

The scientific promise of radio transients is being recognized among the SKA Precursors and Pathfinders. Transients are an explicitly recognized part of the key science case for many instruments (e.g., ASKAP, MeerKAT, LOFAR, MWA) or implicitly by being a significant fraction of the time awarded on the telescopes (e.g., EVLA). A recent example of the continued exploration of the transient sky, with particular relevance to the low-frequency component of the SKA and its pathfinders was a short search campaign carried out at the Long Wavelength Demonstrator Array (LWDA). The LWDA was a 16-dipole, dual-polarization sparse aperture array, operating over the frequency range of 60–80 MHz, and located near the center of the National Radio Astronomy Observatory's Very Large Array (VLA). An all-sky imaging correlator was developed for it, and it took observations over the course of about 6 months. No celestial radio transients were detected [36], but Fig. 1 illustrates the potential of an LWDA-like sparse aperture array for searching for transients.

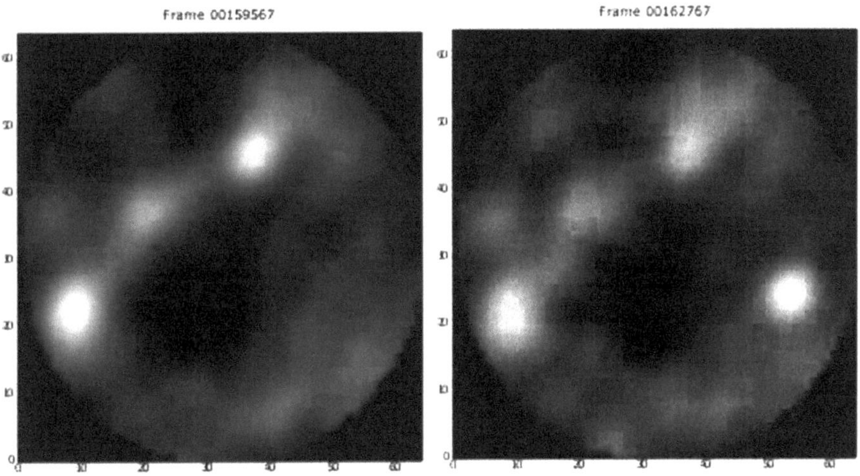

Fig. 1 Terrestrial radio transient captured by the Long Wavelength Demonstrator Array. (*Left*) All-sky LWDA image acquired at 60 MHz. The Galactic plane slopes diagonally from the *upper right* to the *lower left* and the sources Cyg A and Cas A are visible in addition to a general enhancement toward the inner Galaxy. (*Right*) An image acquired only seconds later. The dominant source is the reflection of a TV transmitter located hundreds of kilometers away, which is reflecting off an ionized meteor trail. While this particular transient would represent radio frequency interference (RFI), all-sky imaging at frequencies not used by TV or other transmitters could be a powerful way for aperture arrays to search for radio transients

In summary, the future for radio transients seems promising, both scientifically and technically. Scientifically, there are a host of questions, across the field of astronomy, that can be addressed by opening up the time domain at radio wavelengths. Technically, there are a host of instruments, either beginning to operate or that will soon be operating that will dramatically improve our knowledge of the radio transient sky, leading toward the construction of the SKA.

Acknowledgements I am privileged to have had many enlightening conversations with numerous individuals on these topics. An almost certainly incomplete list includes P. Backus, G. Bower, S. Chatterjee, J. Cordes, P. Dewdney, D. Frail, S. Hyman, M. McLaughlin, N. Kassim, and J. Tarter. This research was carried out at the Jet Propulsion Laboratory, California Institute of Technology, under a contract with the National Aeronautics and Space Administration.

References

1. Abadie, J., et al.: Search for gravitational waves from compact binary coalescence in LIGO and virgo data from S5 and VSR1. Phys. Rev. D. **82**, 102001 (2010)
2. Berger, E.: Flaring up all over-radio activity in rapidly rotating late M and L dwarfs. Astrophys. J. **572**, 503–513 (2002)
3. Berger, E.: Radio observations of a large sample of Late M, L, and T dwarfs: The distribution of magnetic field strengths. Astrophys. J. **648**, 629–636 (2006)
4. Berger, E., Ball, S., Becker, K.M., et al.: Discovery of radio emission from the brown dwarf LP944-20. Nature **410**, 338–340 (2001)
5. Blandford, R.D., et al.: New Worlds, New Horizons in Astronomy and Astrophysics. National Academies Press, Washington, DC (2010)
6. Bloom, J.S., et al.: Coordinated Science in the Gravitational and Electromagnetic Skies. Astro2010 Science White Paper. arXiv:0902.1527
7. Börner, G., Cohen, J.M.: Pulsars—explanation for observed glitches. Nature Phys. Sci. **231**, 146–147 (1971)
8. Bower, G.C., Saul, D., Bloom, J.S., et al: Submillijansky transients in archival radio observations. Astrophys. J. **666**, 346–360 (2007)
9. Burgay, M., D'Amico, N., Possenti, A., et al.: An increased estimate of the merger rate of double neutron stars from observations of a highly relativistic system. Nature **426**, 531–533 (2003)
10. Burke-Spolaor, S., Bailes, M., Ekers, R., Macquart, J.-P., Crawford, F.: III: Radio bursts with extragalactic spectral characteristics show terrestrial origins. Astrophys. J. **727**, 18 (2011)
11. Carilli, C.L., Rawlings, S.: Motivation, key science projects, standards and assumptions. New Astron. Rev. **48**, 979–984 (2004)
12. Cen, R., Ostriker, J.P.: Where are the baryons? Astrophys. J. **514**, 1–6 (1999)
13. Cocconi, G., Morrison, P.: Searching for interstellar communications. Nature **184**, 844–846 (1959)
14. Cognard, I., Shrauner, J.A., Taylor, J.H., Thorsett, S.E.: Giant radio pulses from a millisecond pulsar. Astrophys. J. **457**, L81–L84 (1996)
15. Cordes, J.M., Lazio, T.J.W., McLaughlin, M.A.: The dynamic radio sky. New Astron. Rev. **48**, 1459–1472 (2004)
16. Cordes, J.M., Bhat, N.D.R., Hankins, T.H., McLaughlin, M.A., Kern, J.: The brightest pulses in the universe: Multifrequency observations of the crab pulsar's giant pulses. Astrophys. J. **612**, 375–388 (2003)

17. Cordes, J.M., Lazio, T.J.W., Sagan, C.: Scintillation-induced intermittency in SETI. Astrophys. J. **487**, 782–808 (1997)
18. Davé, R., Hernquist, L., Katz, N., Weinberg, D.H.: The low-redshift lyα forest in cold dark matter cosmologies. Astrophys. J. **511**, 521–545 (1999)
19. Davé, R., Cen, R., Ostriker, J.P., et al.: Baryons in the warm-hot intergalactic medium. Astrophys. J. **552**, 473–483 (2001)
20. de Zeeuw, P.T., Molster, F.J. (eds.): A science vision for european astronomy. ASTRONET (2007) `http://www.astronet-eu.org/`
21. Guedel, M., Benz, A.O.: X-ray/microwave relation of different types of active stars. Astrophys. J. **405**, L63–L66 (1993)
22. Hallinan, G., Bourke, S., Lane, C., et al.: Periodic bursts of coherent radio emission from an ultracool dwarf. Astrophys. J. **663**, L25–L28 (2007)
23. Hallinan, G., Antonova, A., Doyle, J.G., Bourke, S., Lane, C., Golden, A.: Confirmation of the electron cyclotron maser instability as the dominant source of radio emission from very low mass stars and brown dwarfs. Astrophys. J. **684**, 644–653 (2008)
24. Hankins, T.H., Rickett, B.J.: Methods in computational physics. Radio Astron. **14**, 55 (1975)
25. Hansen, B.M.S., Lyutikov, M.: Radio and X-ray signatures of merging neutron stars. Mon. Not. R. Astron. Soc. **322**, 695–701 (2001)
26. Hawking, S.W.: Black hole explosions? Nature **248**, 30–31 (1974)
27. Hewish, A., Bell, S.J., Pilkington, J.D.H., Scott, P.F., Collins, R.A.: Observation of a rapidly pulsating radio source. Nature **217**, 709–713 (1968)
28. Howk, J.C., Ribaudo, J.S., Lehner, N., Prochaska, J.X., Chen, H.-W.: Strong $z \sim 0.5$ O VI absorption towards PKS 0405−123: implications for ionization and metallicity of the cosmic web. Mon. Not. R. Astron. Soc. **396**, 1875–1894 (2009)
29. Hyman, S.D., Lazio, T.J.W., Kassim, N.E., Ray, P.S., Markwardt, C.B., Yusef-Zadeh, F.: A powerful bursting radio source towards the galactic centre. Nature **434**, 50–52 (2005)
30. Jansky, K.G.: Electrical disturbances apparently of extraterrestrial origin. Proc. Inst. Radio Eng. **21**, 1387–1398 (1933)
31. Johnston, S., Romani, R.W.: Giant pulses from PSR B0540−69 in the large magellanic cloud. Astrophys. J. **590**, L95–L98 (2003)
32. Keane, E.F., Kramer, M.: On the birthrates of galactic neutron stars. Mon. Not. R. Astron. Soc. **391**, 2009–2016 (2008)
33. Lazio, J., et al.: Magnetospheric Emissions from Extrasolar Planets. Astro2010: The Astronomy and Astrophysics Decadal Survey, Science White Papers, No. 177 (2009)
34. Lazio, T.J.W., Shankland, P.D., Farrell, W.M., Blank, D.L.: Radio observations of HD 80606 near planetary periastron. Astron. J. **140**, 1929–1933 (2010a)
35. Lazio, T.J.W., Carmichael, S., Clark, J., et al.: A blind search for magnetospheric emissions from planetary companions to nearby solar-type Stars. Astron. J. **139**, 96–101 (2010b)
36. Lazio, T.J.W., Clarke, T.E., Lane, W.M., et al.: Surveying the dynamic radio sky with the long wavelength demonstrator array. Astron. J. **140**, 1995–2006 (2010c)
37. Lorimer, D.R., Bailes, M., McLaughlin, M.A., Narkevic, D.J., Crawford, F.: A bright millisecond radio burst of extragalactic origin. Science **318**, 777–780 (2007)
38. McLaughlin, M.A., Lyne, A.G., Lorimer, D.R., et al.: Transient radio bursts from rotating neutron stars. Nature **439**, 817–820 (2006)
39. McLaughlin, M.A. Cordes, J.M.: Searches for giant pulses from extragalactic pulsars. Astrophys. J. **596**, 982–966 (2003)
40. Ofek, E.O., Breslauer, B., Gal-Yam, A., et al.: Long-duration radio transients lacking optical counterparts are possibly galactic neutron stars. Astrophys. J. **711**, 517–531 (2010)
41. Osullivan, J.D., Ekers, R.D., Shaver, P.A.: Limits on cosmic radio bursts with microsecond time scales. Nature **276**, 590–591 (1978)
42. Phinney, S., Taylor, J.H.: A sensitive search for radio pulses from primordial black holes and distant supernovae. Nature **277**, 117–118 (1979)
43. Rees, M.J.: A better way of searching for black-hole explosions. Nature **266**, 333–334 (1977)

44. Rickett, B.J.: Interstellar scintillation and pulsar intensity variations. Mon. Not. R. Astron. Soc. **150**, 67 (1970)
45. Romani, R.W., Johnston, S.: Giant pulses from the millisecond pulsar B1821−24. Astrophys. J. **557**, L93–L96 (2001)
46. Savage, B.D., Lehner, N., Wakker, B.P., Sembach, K.R., Tripp, T.M.: Detection of ne VIII in the low-redshift warm-hot intergalactic medium. Astrophys. J. **626**, 776–794 (2005)
47. Scargle, J., Pacini, F.: Pulsar glitches—mechanism for crab nebula pulsar. Nature. Phys. Sci. textbf232, 144–149 (1971)
48. Tarter, J.: The search for extraterrestrial intelligence (SETI). Ann. Rev. Astron. Astrophys. **39**, 511–548 (2001)
49. Turolla, R., Possenti, A., Treves, A.: Is the bursting radio source GCRT J1745-3009 a double neutron star binary? Astrophys. J. **628**, L49–L52 (2005)
50. Weisberg, J.M., Nice, D.J., Taylor, J.H.: Timing measurements of the relativistic binary pulsar PSR B1913+16. Astrophys. J. **722**, 1030–1034 (2010)
51. Zhang, B., Gil, J.: GCRT J1745−3009 as a transient white dwarf pulsar. Astrophys. J. **631**, L143–L146 (2005)

Cosmic Magnetism: Current Status and Outlook to the SKA

Marijke Haverkorn

Abstract This paper describes the role that the Square Kilometre Array can play in studies of Cosmic Magnetism, with emphasis on one of the main tracers of cosmic magnetism, viz radio polarimetry. Radio polarimetric observations allow measurement of various components of cosmic magnetic fields in the diffuse ionized gas, through polarized synchrotron emission, Faraday rotation or both. SKA will produce a Rotation Measure Grid of polarized background sources of unprecedented density and accuracy, which will allow determination of the magnetic field topology in the Milky Way, external galaxies and galaxy clusters out to high redshift. Intergalactic magnetic fields may be detected and studied. The SKA signifies a huge step in cosmic magnetism studies, which will allow us to map out the evolution of galactic magnetic fields to high redshifts, to distinguish between models of the origin and evolution of cosmic magnetism, and to characterize intracluster and intergalactic magnetic fields.

1 Introduction

Magnetism is one of the fundamental forces of physics. Magnetic fields interact with charged particles through the Lorentz force, which makes the particles gyrate around magnetic field lines. As most of the Universe is ionized, this provides a tight feedback between interstellar plasma and magnetic field throughout the Universe. Therefore, magnetic fields have a major influence on many of the physical processes in the Universe (mostly those where gravity is not important). The magnetic field in disk galaxies provides a pressure component comparable to the turbulent gas and

M. Haverkorn (✉)
ASTRON, PO Box 2, 7990 AA, Dwingeloo, The Netherlands

Leiden Observatory, Leiden University, P.O. Box 9513, 2300 RA Leiden, The Netherlands
e-mail: haverkorn@astron.nl

D. Barbosa et al. (eds.), *The Square Kilometre Array: Paving the way for the new 21st century radio astronomy paradigm*, Astrophysics and Space Science Proceedings,
DOI 10.1007/978-3-642-22795-0_7,

cosmic ray pressures. This pressure component contributes significantly to the total pressure which counterbalances gas disks against gravity [13]. Magnetic fields also influence gas flows, e.g. in spiral arms [24], turbulent interstellar plasma [19, 47] or neutral hydrogen clouds [37]. Cosmic rays are also strongly coupled to interstellar magnetic fields through the Lorentz force. The streaming motions of cosmic rays along the field lines excite resonant Alfvén waves which then scatter them (e.g. [33]). It is this scatter of cosmic rays off magnetic field irregularities that accelerates them to their relativistic velocities.

2 Methods to Measure Magnetic Fields

There are many ways to measure magnetic field strength and/or direction in the various components of the interstellar, intergalactic and intracluster medium. One of the main tracers of magnetic fields is radio polarimetry, which allows magnetic field measurements through synchrotron radiation and through Faraday rotation. Zeeman splitting of spectral lines gives the strength of the magnetic field in cold gas clouds [53], external galaxies [46] and starforming regions [56]. Polarized emission from magnetized dust in dense clouds along the direction of the local magnetic field can be observed in the submm or far infrared, while polarization in optical starlight absorbed by magnetized dust grains in diffuse dust clouds indicates the magnetic field direction in those clouds. In this paper, the focus will be on radio polarimetry of synchrotron radiation and Faraday rotation methods.

2.1 Radio Synchrotron Emission

Charged particles moving in the presence of a magnetic field will undergo acceleration through the Lorentz force, and subsequently radiate. If the particles are moving at relativistic speeds, this radiation is synchrotron radiation. The amount of synchrotron emission observed depends on the presence of a magnetic field component perpendicular to the line of sight, $B_\perp$. Optically thin synchrotron emission ϵ_s at frequency ν of an ensemble of relativistic electrons with a power law energy spectrum $N(E)dE \sim E^\gamma dE$ in a uniform magnetic field depends on the magnetic field component in the plane of the sky $B_\perp$ according to

$$\epsilon_s \sim B_\perp^{(\gamma+1)/2} \nu^{-(\gamma-1)/2}. \quad (1)$$

Synchrotron emission provides information about the slope of the energy spectrum of the electrons, and an estimate of the strength of the magnetic field if assumptions are made about the volume of the source and the energy density and composition of cosmic rays.

Synchrotron emission from a region with a uniform magnetic field has a theoretical maximum polarization degree of $\sim$70% [43], with the plane of polarization perpendicular to the direction of $B_{\perp}$ in the plane of the sky. The angle of linear polarization of synchrotron emission therefore gives information on the orientation of the magnetic field component perpendicular to the line of sight, at high frequencies where Faraday rotation can be neglected. In practice, the observed emission is integrated over large regions in space with a complicated magnetic field structure and an ionized plasma present in addition to the relativistic electrons. Consequently, the integrated emission is usually much less polarized than the theoretical 70% (see e.g. [52]).

2.2 *Faraday Rotation of Compact Sources*

Faraday rotation is the rotation of the plane of polarization of a linearly polarized electromagnetic wave, which propagates through a region of free electrons permeated by a magnetic field, such as the interstellar plasma. The linear polarization angle θ depends on the intrinsic polarization angle at emission θ_0, the Rotation Measure (RM) and on the wavelength of the electromagnetic wave λ as

$$\begin{aligned} \theta &= \theta_0 + \mathrm{RM}\lambda^2 \\ &= \theta_0 + \left[0.812 \int \left(\frac{n_e}{\mathrm{cm}^3}\right)\left(\frac{\mathbf{B}}{\mu\mathrm{G}}\right) \cdot \left(\frac{\mathbf{ds}}{\mathrm{pc}}\right)\right]\left(\frac{\lambda}{\mathrm{m}}\right)^2 \quad [\mathrm{rad}] \end{aligned} \tag{2}$$

where n_e is the electron density, $\mathbf{B}$ is the magnetic field, and $d\mathbf{s}$ is the path length through the Faraday-rotating medium.

Therefore, measurements of θ at multiple wavelengths can determine the RM for a given source as the slope of the graph of θ versus λ^2. At least three frequency (or wavelength) channels are needed to properly determine the RM. However, this method only works if the synchrotron emitting and Faraday rotating medium are completely separate, e.g. for radiation from a radio galaxy without thermal gas, which is Faraday rotated by the foreground magneto-ionic medium in the Milky Way, or the RM of a pulsar.

For polarized background extragalactic sources (EGS) or pulsars behind a magneto-ionized foreground, RMs are powerful tools to probe the foreground magnetic field. These measurements are used to probe the magnetic field in the Milky Way (e.g. [11], [41]). Since it is possible to estimate the distance to these sources, and if we make some assumptions about the electron density distribution within the Galaxy, then we can work backwards to estimate what the magnetic field might look like along their particular lines-of-sight by using large-scale magnetic field models.

Subsequently, the goal for such observations is to measure RMs for the highest density of sources possible, allowing for the most accurate reconstruction of the intervening field. Consequently, much effort has been put into obtaining a high spatial and angular density of RM sources, in the Galactic plane (e.g. the Canadian and Southern Galactic Plane Surveys [28, 38, 50]), and across the whole sky (e.g. [51]). This method can be applied to other nearby galaxies as well, as long as they have large enough angular size, or to galaxy clusters (e.g. [10]). At this moment this is done for a limited amount of external galaxies such as the Large Magellanic Cloud [23], the Small Magellanic Cloud [35], and M31 [26]. However, this number will increase greatly with the SKA, see Sect. 4.

2.3 Rotation Measure Synthesis

If the synchrotron emission and Faraday rotation in a plasma are mixed, as expected for most of the interstellar medium, (2) does not hold anymore. Instead, a Fourier relation is needed, which can disentangle various emission components and their partial rotation measure, called Faraday depth. This method is named Rotation Measure synthesis, also sometimes called Faraday Tomography. The idea was described by [15] decades ago, but only now technological development has made spectropolarimetric capabilities high enough to use this technique [14]. The Faraday depth ϕ of a Faraday rotating medium is defined as

$$\phi = 0.81 \int_{x_1}^{x_2} n_e \mathbf{B} \cdot \mathbf{ds},$$

where the integral runs over a particular part of the line of sight from x_1 to x_2. If the synchrotron emitting medium is located behind the Faraday rotating screen, $\phi = \mathrm{RM}$. However, if synchrotron emission and Faraday rotation occur in the same medium and/or alternatingly along the line of sight, partial or total depolarization of the emission occurs. The observed polarized flux density is $\tilde{P}(\lambda^2) = W(\lambda^2)P(\lambda^2)$, where $P(\lambda^2)$ is the complex polarized surface brightness as a function of wavelength and $W(\lambda^2)$ is a weight function or sampling function which is non-zero only at the wavelengths where measurements are taken. Fourier transformation of the observed polarized surface brightness as a function of wavelength results in $F(\phi)$, the complex polarized surface brightness per unit Faraday depth ϕ as

$$P(\lambda^2) = \int_{-\infty}^{\infty} F(\phi) \mathrm{e}^{2i\phi\lambda^2} d\phi.$$

Thus RM synthesis provides a three-dimensional map of polarized intensity as a function of Faraday depth. However, as Faraday depth or RM does not necessarily increase monotonically with distance (i.e. significant changes in field direction, electron density, or magnetic field strength will also impact the RM value), there is no direct distance information in these maps. Indirect distance information can be obtained from morphological correspondences with structures such as spiral arms, supernova remnants or external galaxies. Since the traditional way of measuring RMs only provided one RM value for each line of sight, and none where depolarization was severe, this new technique provides a tremendous increase in the amount of obtainable information from radio spectropolarimetric surveys.

3 The Role of Magnetic Fields in Interstellar Gas

The strength of the influence of magnetic fields on the interstellar medium (ISM) of galaxies is commonly quantified by comparing energy densities or pressures of the various components. Figure 1 shows energy density estimates in the Galaxy as a function of distance from the Galactic center [29]. The magnetic field energy density $|B|^2/8\pi$ is based on equipartition values derived from radial profiles as modeled by [9]. This method was used by Berkhuijsen in [4] to obtain the Galactic magnetic field strength as a function of Galactocentric radius. The figure also shows the difference between the classical equipartition formula and the revised formula based on integration over a fixed energy range instead of a frequency range [8]. The thermal gas energy density is only indicated at the solar radius and is based on standard values of the densities and temperatures of the cold, warm and hot gas components at the solar radius. The turbulent gas energy density ρv_{turb}^2 is calculated using an exponential gas scale length of 3.15 kpc and a turbulent velocity based on [36]. They found that the velocity fluctuations in the (inner) Galaxy are best described by three components: a cold component with a velocity dispersion $\Delta v = 6.3\,\mathrm{km\,s^{-1}}$, a warm component with $\Delta v = 12.3\,\mathrm{km\,s^{-1}}$, and a fast component of $\Delta v = 25.9\,\mathrm{km\,s^{-1}}$, probably related to large-scale motions. All components

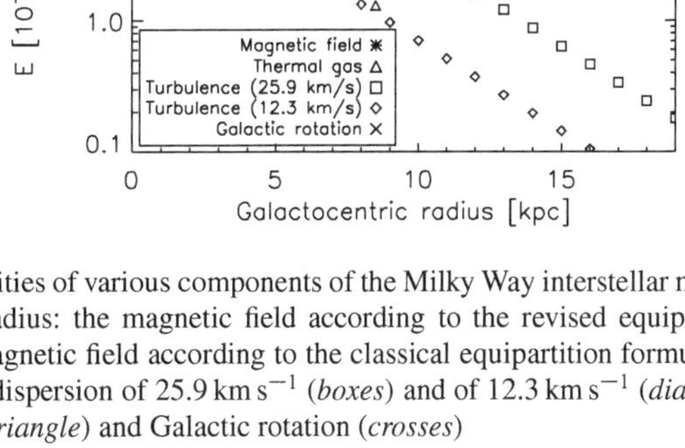

Fig. 1 Energy densities of various components of the Milky Way interstellar medium as a function of Galactocentric radius: the magnetic field according to the revised equipartition function [8] (*large asterisks*), magnetic field according to the classical equipartition formula (*small asterisks*), gas with a velocity dispersion of 25.9 km s^{-1} (*boxes*) and of 12.3 km s^{-1} (*diamonds*), thermal gas at the solar radius (*triangle*) and Galactic rotation (*crosses*)

are more or less constant with radius. Both the warm turbulent component with $\Delta v = 12.3\,\mathrm{km\,s^{-1}}$ and the large-scale motion component with $\Delta v = 25.9\,\mathrm{km\,s^{-1}}$ are shown in the figure. Throughout, a solar radius of 8.5 kpc is used.

Figure 1 shows that the magnetic and turbulent energy densities are, given the uncertainties in the assumptions made, similar, while the energy density in the thermal gas components is much less. The dominance of magnetic pressure over the thermal pressure is consistent with the observed remarkable uniformity in magnetic field strength over a wide range of gas densities [54]. In a similar analysis to Fig. 1, [3] finds for external spiral galaxy NGC 6946 equipartition between magnetic fields and turbulent gas density as well, including a possible dominance of magnetic fields over turbulence in the outer regions of the galaxy.

4 Observing Cosmic Magnetism with the SKA

The SKA will bring a major step in analysis of cosmic magnetism, for a large part through measurements of Faraday rotation of background sources in a so-called *RM Grid*, and through RM Synthesis analysis.

Currently, the Galactic plane is covered with RM Grids with a density of about one source per square degree [11, 12]. The Northern sky is covered with a grid of similar density [51] using more than 35,000 polarized sources in the northern sky from the NRAO VLA Sky Survey (NVSS, [17]), but since this survey only consists of two frequency bands, care should be taken in interpretation of high RMs from this survey.

This situation can be immensely improved with the Square Kilometre Array. [7] calculate the expected detections of background source RMs for an SKA all-sky survey at 21 cm with an integration time of one hour per pointing, giving a sensitivity in linear polarization of $\sim 0.1\,\mu$Jy. They predict to find 2×10^7 RMs over the whole sky, at a mean separation of $\sim 90''$ between adjacent measurements. Figure 2 illustrates the incredible improvement over the currently available data. Continuum radio emission at 20 cm is shown in grey scale, with RMs of 21 polarized background sources, shown inside the large red circles in the [26]. The light-red dots covering the whole map show the improvement that the SKA will provide. It is clear that the magnetic field in the nearby galaxy M31 can be probed in great detail using the RM Grid as provided by the SKA.

A radio polarimetric survey with SKA, needed to attain these results, should have high polarization purity, excellent uv-coverage and significant collecting area in the core to enable imaging of diffuse emission, and a broad and fairly regular frequency coverage for accurate RM Synthesis. Therefore, an important requirement for the SKA for cosmic magnetism research will be spectropolarimetric capability.

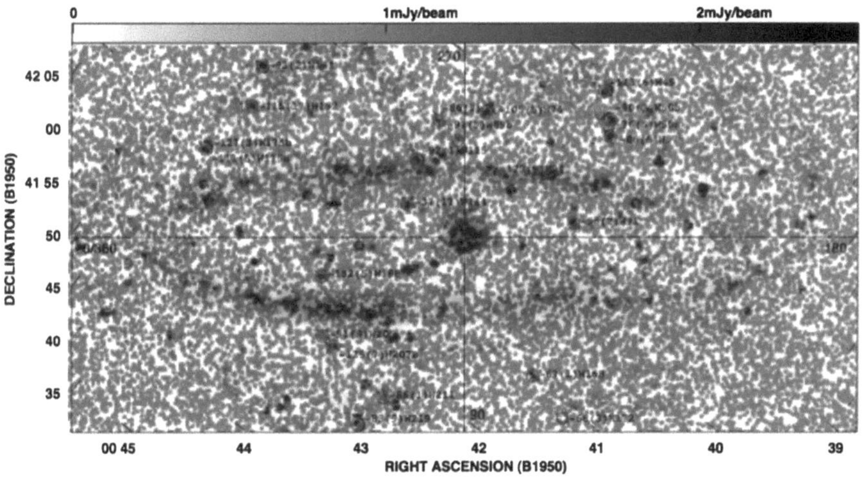

Fig. 2 Grey scale: 20 cm radio continuum emission of M31 [5], where the 21 *large red circles* indicate background point sources with measured rotation measures from [26]. The overplotted *light-red dots* show the number of background source RMs that are expected to be detected with the SKA

5 Cosmic Magnetism Questions to be Addressed with the SKA

5.1 Interstellar Magnetic Fields

In the plane of the Milky Way, the magnetic field is oriented parallel to the plane on large scales, as is evident from starlight polarization [21] and high-frequency (submm) polarization maps (e.g. [44]). The total magnetic field strength in the solar neighborhood is about 6 μG, mostly dominated by a small-scale, turbulent field. Any vertical component of the magnetic field at the Sun is small, if present at all. [34] find a small coherent magnetic field component towards the Southern Galactic pole of $B_z \approx 0.3\,\mu$G, but find no evidence for a coherent vertical field towards the Northern Galactic pole.

One large-scale reversal in the azimuthal direction of the magnetic field in the plane has been known for decades, but the exact number and location of reversals is still under debate: models of the large-scale magnetic field in the Milky Way can have widely different conclusions based on similar data sets [25, 55]. This discrepancy is due to small-scale field components unaccounted for in the models, sparse sampling of background sources and the lack of sensitivity to detect pulsars at larger distances from the sun. An SKA RM Grid will vastly improve the numbers of data points, allowing direct inversion of the data distribution to a magnetic field geometry using wavelet transforms [48].

Nearby spiral galaxies host similar magnetic field patterns, with total field strengths between 5 and 30 μG or even somewhat higher for starburst galaxies [3]. Polarized synchrotron emission from halos of edge-on nearby spirals generally show X-shaped patterns, probably due to outflows of gas [31].

With the SKA, the RM Grid analysis will become possible for a large number of external galaxies. Current studies report a few to a few dozen extragalactic sources behind nearby galaxies [23, 26, 35], but deep SKA observations could result in $>10^5$ background sources behind these galaxies. Even a pointing on a spiral galaxy at a distance of 10 Mpc will still include $\sim$50 background sources with which its magnetic field geometry can be determined [7].

The small-scale, turbulent magnetic field component can be traced mostly by using power spectra and structure functions. Evidence for a Kolmogorov spectrum on scales from AU to tens of parsecs has been found [2], however, deviations from that law are observed as well (e.g. [39]). The SKA RM Grid and RM Synthesis of the diffuse emission will make it possible to follow this turbulent behavior to much smaller scales. Studies of turbulence properties through RMs of background sources show structure on scales of degrees and larger with a definite dependence on location in the Galaxy [27, 49]. The SKA will expand parameter space to scales much below an arcminute, and will be able to characterize turbulence in shocks, gas clouds, loops, filaments, etc.

5.2 Intracluster and Intergalactic Magnetic Fields

Many galaxy clusters emit diffuse radio synchrotron emission; either from the radio halo, which is mostly due to turbulent magnetic fields and therefore highly depolarized; or from radio relics off-center to the cluster, thought to be generated by compressed magnetic fields in merger shocks and thus fairly highly polarized [20, 57]. With the SKA, it will become possible to detect magnetic fields in galaxy clusters out to high redshifts, using Rotation Measures of background sources [30].

Detection of intergalactic magnetic fields has been attempted in several ways. [58] correlated all-sky maps of Faraday rotation with the 2MASS and CfA surveys of galaxies. They found a correlation of Faraday rotation with the intergalactic filament in the Perseus–Pisces supercluster and derived an intergalactic magnetic field strength of $0.01 - 0.1\ \mu$G from that. [32] observed extended synchrotron emission in the direction of the Coma cluster, which would correspond to an intergalactic magnetic field strength of 0.1 μG. However, these methods are inherently plagued by uncertainties in subtraction of the Galactic synchrotron foreground.

The second method uses gamma-ray emission from bursting events and their interaction with background photons [45]. For a sufficiently strong intergalactic magnetic field strength, one would observe a halo of gamma-ray photons around the source. Ando and Kusenko [1] stacked the 170 brightest Active Galactic Nuclei from Fermi/LAT and detected a faint gamma-ray halo around the stacked source image, resulting in an intergalactic magnetic field strength of $\sim 10^{-15}$ G. However,

[40] respond that this detection is due to the point spread function of Fermi/LAT. The non-detection of high-energy photons from blazars can be used to estimate that the intergalactic magnetic field fills at least 60% of space with fields stronger than $\sim 10^{-16} - 10^{-15}$ G [18].

With the SKA RM Grid, it may become possible to determine a trend in RM as a function of redshift [22]. This can be used to distinguish between RMs resulting from magnetic fields in the high-redshift sources themselves and those produced by fields in any foreground Lyα absorbers. In this way, the evolution of galactic magnetism can be traced. Currently, this method has been attempted (e.g. [42]), but sparse sampling, low sensitivity, and problems with foreground subtraction preclude clear conclusions.

5.3 Explore the Origin and Evolution of Cosmic Magnetism

Many galaxies seem to exhibit large-scale magnetic field components: spiral magnetic field configurations have been observed in all spiral galaxies [4], but also in flocculent or ring galaxies [16]. This indicates the presence of an agent which organizes the magnetic field and maintains it: a dynamo. Even though dynamo action seems to be ubiquitous, its physics is not well understood. Dynamo models predict that various dynamo modes can be excited [6]. However, data is too scarce at the moment to confidently distinguish between these modes for the most nearby galaxies, or to observe them at all in other galaxies. The SKA will be able to determine the Fourier spectrum of dynamo modes in galaxies out to a distance of 40 Mpc. Therefore, the SKA has the potential to increase the galaxy sample with well-known field patterns by up to three orders of magnitude [7] and finally shed light on the origin and evolution of magnetism in the cosmos.

Acknowledgements The author wishes to express thanks to the Scientific and Local Organizing Committees for their invitation, and for a stimulating and productive meeting.

References

1. Ando, S., Kusenko, A.: Astrophys. J. **722**, 39 (2010)
2. Armstrong, J.W., Rickett, B.J., Spangler, S.R.: Astrophys. J. **443**, 209 (1995)
3. Beck, R.: Astrophys. Space. Sci. **289**, 293 (2004)
4. Beck, R.: Space Sci. Rev. **99**, 243 (2001)
5. Beck, R., Berkhuijsen, E.M., Hoernes, P.: Astron. Astrophys. Suppl. **129**, 329 (1998)
6. Beck, R., Brandenburg, A., Moss, D., Shukurov, A., Sokoloff, D.D.: Annu. Rev. Astron. Astrophys. **34**, 155 (1996)
7. Beck, R., Gaensler, B.M.: New Astron. Rev. **48**, 1289 (2004)
8. Beck, R., Krause, M.: Astron. Nachr. **326**, 414 (2005)
9. Beuermann, K., Kanbach, G., Berkhuijsen, E.M.: Astron. Astrophys. **153**, 17 (1985)

10. Bonafede, A., Feretti, L., Murgia, M., Govoni, F., Giovannini, G., Dallacasa, D., Dolag, K., Taylor, G.B.: Astron. Astrophys. **513**, 30 (2010)
11. Brown, J.C., Haverkorn, M., Gaensler, B.M., Taylor, A.R., Bizunok, N.S., McClure-Griffiths, N.M., Dickey, J.M., Green, A.J.: Astrophys. J. **663**, 258 (2007)
12. Brown, J.C., Taylor, A.R., Jackel, B.J.: Astrophys. J. Suppl. **145**, 213 (2003)
13. Boulares, A., Cox, D.P.: Astrophys. J. **365**, 544 (1990)
14. Brentjens, M.A., de Bruyn, A.G.: Astron. Astrophys. **441**, 1217 (2005)
15. Burn, B.J.: Mon. Not. R. Astron. Soc. **133**, 67 (1966)
16. Chyży, K.T., Buta, R.J.: Astrophys. J. **677**, 17 (2008)
17. Condon, J.J., Cotton, W.D., Greisen, E.W., Yin, Q.F., Perley, R.A., Taylor, G.B., Broderick, J.J.: Astron. J. **115**, 1693 (1998)
18. Dolag, K., Kachelriess, M., Ostapchenko, S., Tomàs, S.R.: Astrophys. J. Lett. **727**, 4
19. Elmegreen, B.G., Scalo, J.: Annu. Rev. Astron. Astrophys. **42**, 211 (2004)
20. Ensslin, T.A., Biermann, P.L., Klein, U., Kohle, S.: Astron. Astrophys. **332**, 395 (1998)
21. Fosalba, P., Lazarian, A., Prunet, S., Tauber, J.A.: Astrophys. J. **564**, 762 (2002)
22. Gaensler, B.M., Beck, R., Feretti, L.: New Astron. Rev. **48**, 1003 (2004)
23. Gaensler, B.M., Haverkorn, M., Staveley-Smith, L., Dickey, J.M., McClure-Griffiths, N.M., Dickel, J.R., Wolleben, M.: Science **307**, 1610 (2005)
24. Gómez, G.C., Cox, D.P.: Astrophys. J. **580**, 235 (2002)
25. Han, J.L.: Chin. J. Astron. Astrophys. Suppl. **6**, 211 (2006)
26. Han, J.L., Beck, R., Berkhuijsen, E.M.: Astron. Astrophys. **335**, 1117 (1998)
27. Haverkorn, M., Brown, J.C., Gaensler, B.M., McClure-Griffiths, N.M.: Astrophys. J. 680, 362 (2008)
28. Haverkorn, M., Gaensler, B.M., McClure-Griffiths, N.M., Dickey, J.M., Green, A.J.: Astrophys. J. Suppl. **167**, 230 (2006)
29. Heiles, C.E., Haverkorn, M.: Space Sci. Rev. in press (2011)
30. Huarte-Espinosa, M., Alexander, P., Bolton, R., Geisbuesch, J., Krause, M.: In: Esquivel A., et al. (eds.) Magnetic Fields in the Universe II: From Laboratory and Stars to the Primordial Universe, Revista Mexicana de Astronomía y Astrofía (Serie de Conferencias), vol. 36, p. 231 (2009)
31. Krause, M., Wielebinski, R., Dumke, M.: Astron. Astrophys. **448**, 133 (2006)
32. Kronberg, P.P., Kothes, R., Salter, C.J., Perillat, P.: Astrophys. J. **659**, 267
33. Kulsrud, R., Pearce, W.P.: Astrophys. J. **156**, 445 (1969)
34. Mao, S.A., Gaensler, B.M., Haverkorn, M., Zweibel, E.G., Madsen, G.J., McClure-Griffiths, N.M., Shukurov, A., Kronberg, P.P.: Astrophys. J. **714**, 1170 (2010)
35. Mao, S.A., Gaensler, B.M., Stanimirovic, S., Haverkorn, M., McClure-Griffiths, N.M., Staveley-Smith, L., Dickey, J.M.: Astrophys. J. **688**, 1029 (2008)
36. McClure-Griffiths, N.M., Dickey, J.M.: Astrophys. J. **671**, 427 (2007)
37. McClure-Griffiths, N.M., Dickey, J.M., Gaensler, B.M., Green, A.J., Haverkorn, M.: Astrophys. J. **652**, 1339 (2006)
38. McClure-Griffiths, N.M., Dickey, J.M., Gaensler, B.M., Green, A.J., Haverkorn, M., Strasser, S.: Astrophys. J. Suppl. **158**, 178 (2005)
39. Minter, A.H., Spangler, S.R.: Astrophys. J. **458**, 194 (1996)
40. Neronov, A., Semikoz, D.V., Tinyakov, P.G., Tkachev, I.I.: Astron. Astrophys. **526**, 90
41. Nota, T., Katgert, P.: Astron. Astrophys. **513**, 65 (2010)
42. Oren, A.L., Wolfe, A.M.: Astrophys. J. **445**, 624
43. Pacholczyk, A.B.: Radio Astrophysics. Freeman, San Francisco (1970)
44. Page, L., Hinshaw, G., Komatsu, E., Nolta, M.R., Spergel, D.N., et al.: Astrophys. J. Suppl. **170**, 335 (2007)
45. Plaga, R.: Nature **374**, 430 (1995)
46. Robishaw, T., Quataert, E., Heiles, C.E.: Astrophys. J. **680**, 981 (2008)
47. Scalo, J., Elmegreen, B.G.: Annu. Rev. Astron. Astrophys. **42**, 275 (2004)
48. Stepanov, R., Frick, P., Shukurov, A., Sokoloff, D.D.: Astron. Astrophys. 391, 361 (2002)
49. Stil, J.M., Taylor, A.R., Sunstrum, C.: Astrophys. J. **726**, 4 (2011)

50. Taylor, A.R., Gibson, S.J., Peracaula, M., Martin, P.C., Landecker, T.L., et al.: Astron. J. **125**, 3145 (2003)
51. Taylor, A.R., Stil, J.M., Sunstrum, C.: Astrophys. J. **702**, 1230 (2009)
52. Tribble, P.C.: Mon. Not. R. Astron. Soc. **250**, 726 (1991)
53. Heiles, C.E., Troland, T.H.: Astrophys. J. **624**, 773
54. Troland, T.H., Heiles, C.: Astrophys. J. **301**, 339
55. Van Eck, C.L., Brown, J.C., Stil, J.M., Rae, K., Mao, S.A., Gaensler, B.M., Shukurov, A., Taylor, A.R., Haverkorn, M., Kronberg, P.P., McClure-Griffiths, N.M.: Astrophys. J. **728**, 97 (2011)
56. Vlemmings, W.H.T., Diamond, P.J., van Langevelde, H.J., Torrelles, J.M.: Astron. Astrophys. **448**, 597 (2006)
57. van Weeren, R., Röttgering, H.J.A., Brüggen, M., Hoeft, M.: Science **330**, 347 (2010)
58. Xu, Y., Kronberg, P.P., Habib, S., Dufton, Q.W.: Astrophys. J. **637**, 19 (2006)

The SKA and "High-Resolution" Science

A.P. Lobanov

Abstract "High-resolution", or "long-baseline", science with the SKA and its precursors covers a broad range of topics in astrophysics. In several research areas, the coupling between improved brightness sensitivity of the SKA and a sub-arcsecond resolution would uncover truly unique avenues and opportunities for studying extreme states of matter, vicinity of compact relativistic objects, and complex processes in astrophysical plasmas. At the same time, long baselines would secure excellent positional and astrometric measurements with the SKA and critically enhance SKA image fidelity at all scales. The latter aspect may also have a substantial impact on the survey speed of the SKA, thus affecting several key science projects of the instrument.

1 Introduction

The benchmark design for the SKA Phase 1 [35], envisaging operations in the 0.3–10 GHz range and on baselines of up to several hundred kilometres, would have enabled addressing a range of scientific areas relying on sub-arcsecond resolution, including astrometry, pulsar proper motions, supernovae, astrophysical masers, nuclear regions of AGN, physics of relativistic and mildly relativistic outflows, kinetic feedback from AGN, evolution of supermassive black holes and their host galaxies [7]. The revised specifications for the SKA_1 [11, 14], shifting the operational frequency range to 0.07–3 GHz and limiting the baseline length to 100 km, leads to a reduction of the instrumental resolution to $0\overset{''}{.}3$–$1\overset{''}{.}4$ for the dishes (in the 0.45–3 GHz range) and $1''$–$8''$ for the aperture array (in the 0.07–0.45 GHz range).

A.P. Lobanov (✉)
Max-Planck-Institut für Radioastronomie, Auf dem Hügel 69, 53121 Bonn, Germany
e-mail: alobanov@mpifr.de

D. Barbosa et al. (eds.), *The Square Kilometre Array: Paving the way for the new 21st century radio astronomy paradigm*, Astrophysics and Space Science Proceedings,
DOI 10.1007/978-3-642-22795-0_8,

The "stand alone" resolution of SKA_1 will therefore be sufficient for addressing only a subset of topics listed above. Achieving a higher resolution would rely on inclusion of external antennas and operating in the VLBI mode. This would be mostly feasible for the dish part of SKA_1, as most of the present day VLBI arrays are operating at frequencies above 600 MHz, and there are no definite plans to extend VLBI operations to below 300 MHz. SKA_1 operating in Australia can be an integral part of the LBA/NZ Network and EAVN. It would also have a somewhat limited common visibility with the VLBA. SKA_1 sited in South Africa will be a natural partner to EVN+ antennas. In both cases, collaboration with geodetic VLBI is possible, if 2.3 GHz will be maintained as a network frequency by the IVS.

In addition, the sub-arcsecond resolution of SKA_1 may actually be an essential requirement also for achieving the specifications envisaged for the traditional "low-resolution" science, including the surveying capabilities of the array (often viewed as a backbone of the instrument).

These two aspects of the relevance of long baselines to achieving the scientific goals of SKA_1 are discussed below, with Sect. 2 describing potential areas of broad scientific impact of high-resolution studies with SKA_1 and SKA Precursors and Sect. 3 discussing the effect of long baselines on the quality of imaging and surveying capability of SKA_1.

2 High-Resolution Science with SKA and SKA Precursors

With its present design specification, SKA_1 will be most effective addressing the following subset of topics mentioned above: studies of galactic and extragalactic supernova remnants, detection of atomic and molecular gas in galaxies, localisation of non-thermal continuum production sites in AGN, investigations of extragalactic outflows and their role in AGN feedback, studies of supermassive black holes and their relation to galaxy evolution, and research in AGN relic activity.

2.1 Supernova Science

High-sensitivity radio observations at a sub-arcsecond resolution provide an effective tool to detect and monitor extragalactic SN/SNR [3, 25, 27], giving a good estimate of star-formation rate in target galaxies and helping assess the connection between AGN and star formation. These measurements rely on highly sensitive long baselines, and the improvements in sensitivity already enable detecting and imaging much weaker supernova remnants in the Milky Way as well [4, 29]. SKA_1 will enable detecting weaker SN/SNR in a much wider range of galaxies. The combination of high resolution and superb brightness sensitivity would enable much longer tracing of evolving supernova shells and supernova remnants, yielding essential information about their ages and galactic environment.

2.2 Atomic and Molecular Gas in Galaxies

In addition to observations of line emission from H I (at 1.42 GHz) and D I (at 0.327 GHz), studies of OH megamasers (at 1.67 GHz) and H I/OH absorption made with SKA_1 will provide a wealth of information about atomic and molecular gas in the nuclear regions of galaxies [7, 22].

Extragalactic OH megamasers are detected towards IR luminous galaxies and they are 10^3–10^6 times stronger than brightest Galactic masers [16], and they are reported to have a two-component structure [17] possibly tracing an interaction between the ionisation cones of the nuclear outflow and the molecular torus. Combining SKA_1 with antennas on 3000+ km baselines would broaden spectacularly the scope of OH megamaser studies.

Absorption due to several species, most notably H I and OH toward compact continuum sources is a unique tool to probe nuclear regions on parsec scales – still beating the resolution and accuracy of optical integral field spectroscopy studies [31, 33]. In extragalactic objects, OH absorption has been used to probe the conditions in warm neutral gas [16], and CO and H I absorption has become a tool of choice to study the molecular tori [34] and interactions between outflows and ambient ISM [30, 33]. Studies of the nuclear absorption will benefit enormously from highly sensitive baselines provided by using SKA_1 alone or together with VLBI arrays.

2.3 Localisation of Non-thermal Continuum in AGN

Detailed knowledge of the mechanism of high-energy particle and emission production in AGN is pivotal for studies of galactic activity [24] as well as for understanding the kinetic and radiation feedback from AGN influencing cosmological galaxy evolution, black hole growth, and large-scale structure formation.

Self-consistent physical models for non-thermal continuum production in AGN require accurate information about the sites where the bulk of the high-energy emission is produced. Present arguments place these sites anywhere between ~1000 gravitational radii of the central black hole [1] and a ~100 pc separation from it [9].

Joint modelling of radio, optical, and X-ray data in 3C120 indicates that radio flares and X-ray dips seem to originate near the accretion disk [8, 26], while optical and high-energy flares are produced in stationary shocks located at ~1 pc downstream in the jet [2, 20, 36]. One important conclusion from this work is that instantaneous SEDs are likely to result from several physically different plasma components, which may lead to considerable difficulties in their interpretation being made without any reference to spatial location of the emission observed in different bands.

The combination of broad-band monitoring and high-resolution radio observations remains the only viable tool for spatial localisation of non-thermal continuum in AGN. Substantial improvements of sensitivity and time coverage of such combined programs can be achieved with SKA_1 working as part of VLBI experiments, and it would be certainly provide essential information for understanding the physics of high-energy emission production in AGN.

2.4 *Outflows and Feedback in AGN*

Evidence is abound for feedback from AGN to play an important role in physical processes at intergalactic and intracluster scales [5, 32], with the efficiency and mechanism of the kinetic feedback from nuclear outflows still being poorly understood.

SKA_1 will be an excellent tool for probing physical conditions in low-energy tail of outflowing plasma which is believed to carry the bulk of kinetic power of the outflow. This will enable making detailed quantitative studies of evolution and re-acceleration of non-thermal plasma in cosmic objects and provide essential clues for understanding the power and efficiency of the kinetic feedback from AGN and its effect on activity cycles in galaxies and cosmological growth of supermassive black holes. Such studies are critically needed for making a detailed assessment of the role played by AGN in the formation and evolution of the large-scale structure in the Universe.

2.5 *Galactic Mergers and Supermassive Black Holes*

High-resolution and high-sensitivity radio observations are expected to provide arguably the best AGN and SMBH census up to very high redshifts [7]. This will enable cosmological studies of SMBH growth, galaxy evolution, and the role played by galactic mergers in nuclear activity and SMBH evolution.

Most powerful AGN are produced by galactic/SMBH mergers [12, 15]. Activity is reduced when a loss cone is formed and most of nuclear gas is accreted onto SMBH [28]. The remaining secondary SMBH helps maintaining activity of the primary [13], and the evolution of nuclear activity can be connected to the dynamic evolution of binary SMBH in galactic centres [23]. Direct detections of secondary SMBH in post-merger galaxies are the best way to the evolution of black holes and galaxies together. Some of the secondary BH may be "disguised" as ULX objects [29] accreting at 10–5 of the Eddington rate. They are not detected in deep radio images at present.

SKA_1 would be a superb tool for detecting and classifying such objects, thus providing an essential observational information about the SMBH evolution in post-merger galaxies and its influence on the galactic activity, formation of collimated outflows and feedback from AGN.

2.6 *Radio Relics and AGN Cycles*

Nuclear activity in galaxies is believed to be episodic or intermittent, with estimates of activity cycles reaching up to 10^8 years [18, 38]. The episodes of activity are related closely to mergers of galaxies [37] and evolution of supermassive binary black holes resulting from galactic mergers [23, 28]. Both the onset and the latest stages of the jet activity are poorly studied at the moment, because in either case the flowing plasma emit largely at low frequencies.

Relics of previous cycles of nuclear activity are difficult to detect at centimetre wavelengths because of significant losses due to expansion and synchrotron emission. At centimetre wavelengths, such relics decay below the sensitivity limits of the present-day facilities within 10^4–10^5 years after the fuelling of extended lobes stops. This explains the relatively small number of such relics known so far. SKA_1, working below 1 GHz, would be able to detect such relics for at least 10^7 years after the fuelling stops, and this would make it possible to assess the activity cycles in a large number of objects, searching for signs of re-started activity in radio-loud objects and investigating "paleo" activity in presently radio-quiet objects. This information will be essential for constructing much more detailed models of evolution and nuclear activity of galaxies.

3 Long Baselines for "Low Resolution" Science

Availability of long baselines is not only a definitive requirement for a number of science areas relying on high-resolution imaging, but also an important factor for achieving design specifications envisaged for several key science areas of SKA_1 and full SKA. This concerns primarily the design goals for the survey speed and r.m.s. sensitivity to extended, low-surface emission.

3.1 *Imaging with the SKA*

In the SKA developments, imaging capability is often viewed as a "tradeoff" against the survey speed [10] – hence "core-spread" array configurations are favoured and long baselines are downgraded [11, 14]. But the two-order of magnitude sensitivity improvement envisaged for SKA will lead to situation when surveying is made in crowded fields with a number of resolved objects per primary beam of the receiving element.

In this circumstance, high fidelity imaging becomes then an essential feature rather than a "tradeoff", and the combination $A_{\rm eff}/T_{\rm sys}$ becomes inadequate as a figure of merit describing the r.m.s. sensitivity and survey speed, $S_{\rm s} \propto (A_{\rm eff}/T_{\rm sys})^2$.

It has therefore been argued [6, 21] that SKA needs to have the capability of imaging adequately (at least) all those spatial frequencies at which there is more than one sky object per primary beam.

3.2 Survey Speed Versus Imaging Fidelity

Optimisation for the survey speed as expressed by the $A_{\mathrm{eff}}/T_{\mathrm{sys}}$ factor is based on two implicit assumptions:

(1) The number of objects N_{source} detected in the primary beam area is small, $N_{\mathrm{source}} \le (B_{\mathrm{max}}/D_{\mathrm{ant}})^2$ (where D_{ant} is the diameter of antenna and B_{max} is the maximum baseline length);
(2) All these objects are essentially unresolved by the array, implying that they have angular sizes $\theta_{\mathrm{source}} \le 2\,\mathrm{FWHM}/\sqrt{\mathrm{SNR}}$.

Neither of these two conditions will be realised in the case the case of both SKA and SKA_1. The primary beam area of a 12-m dish will contain ~50 objects ($S_{1.4\mathrm{GHz}} > 0.4\,\mathrm{mJy}$) larger than θ_{source}. This leads to a strong requirement of optimisation for dynamic range and high-fidelity imaging (hence the distribution of the collecting area).

This effect has been investigated in a number of studies [6, 19, 21] showing that a uniform structural sensitivity is desired to alleviate the problems posed by resolved objects in the field of view. The structural sensitivity can described by the *uv*-gap parameter [21], $\Delta u/u$. For an array uniformly sensitive to all detectable angular scales, $\Delta u/u = const$ on all baselines. The structural sensitivity of an array can be described by a factor

$$\eta_{uv} = \exp\left[\frac{\pi^2}{16\ln 2}\frac{\Delta u}{u}\left(\frac{\Delta u}{u}+2\right)\right]^{-1},$$

with $\eta_{uv} \equiv 1$ for a filled aperture (for which $\Delta u/u \equiv 0$). An imperfect *uv*-coverage of an interferometric array results in an additional, scale-dependent factor of the image noise, $\sigma_{uv} = \sigma_{\mathrm{rms}}/\eta_{uv}$, and this has to be taken into account when estimating the performance of SKA for surveys. These estimates should be done using the factor $\eta_{uv}(A_{\mathrm{eff}}/T_{\mathrm{sys}})$.

3.3 Configuration Choices

Analysis based on the structural sensitivity factor shows that, compared to the "core-spread" array configuration (adopted as a benchmark for SKA_1 and SKA, a logarithmic [6, 21] or gaussian array of the same extent in baseline length will have an approximately three times worse brightness sensitivity on scales of $\le 0.1\,B_{\mathrm{max}}$, a ~30% lower noise in snapshot images, and a ~2 times lower sidelobe level. For a ten times larger gaussian/logarithmic array, these figures are reduced only slightly, while the confusion limit is lowered by a factor of 100. These arguments show that adopting such a configuration is rather a necessity for large area surveys. In contrast, for the core-spread configuration, the noise increase due to σ_{uv} would have to be offset by increasing the observing by a factor of 10(!) if one would require to reach within 5% of the r.m.s. specification based on the simple $A_{\text{eff}}/T_{\text{sys}}$ factor.

4 Conclusions

SKA_1 would be able to address a number of important astrophysical areas of study relying on high-resolution radio observations: studying supernovae; providing a good account of starburst activity in galaxies; using megamasers and nuclear absorption to probe the nuclear gas in galaxies; understanding in detail the physics of (ultra- and mildly-relativistic) outflows and their connection to the nuclear regions in galaxies; searching for radio emission from weaker AGN and secondary black holes in post-merger galaxies; and addressing the questions of relic activity and activity cycles in AGN. Reliable high resolution imaging is also needed for achieving the scientific goals in the key science areas of SKA_1 requiring low-resolution imaging and surveying of large areas in the sky, most notably the science areas considered as the very "selling point" of the instrument.

References

1. Acciari, V.A., Aliu, E., Arlen, T., Bautista, M., Beilicke, M., Benbow, W., Bradbury, S.M., Buckley, J.H., Bugaev, V., Butt, Y., et al.: Radio imaging of the very-high-energy γ-ray emission region in the central engine of a radio galaxy. Science **325**, 444 (2009)
2. Arshakian, T.G., León-Tavares, J., Lobanov, A.P., Chavushyan, V.H., Shapovalova, A.I., Burenkov, A.N., Zensus, J.A.: Observational evidence for the link between the variable optical continuum and the subparsec-scale jet of the radio galaxy 3C 390.3. Mon. Not. R. Astron. Soc. **401**, 1231 (2010)
3. Beswick, R.J., Riley, J.D., Marti-Vidal, I., Pedlar, A., Muxlow, T.W.B., McDonald, A.R., Wills, K.A., Fenech, D., Argo, M.K.: 15 years of very long baseline interferometry observations of two compact radio sources in Messier 82. Mon. Not. R. Astron. Soc. **369**, 1221 (2006)
4. Bietenholz, M.F., Bartel, N., Milisavljevic, D., Fesen, R.A., Challis, P., Kirshner, R.P.: The first VLBI image of the young, oxygen-rich supernova remnant in NGC 4449. Mon. Not. R. Astron. Soc. **409**, 1594 (2010)
5. Binney, J., Tabor, G.: Evolving cooling flows. Mon. Not. R. Astron. Soc. **276**, 663 (1995)
6. Bunton, J.: Area Scaling for the SKA. SKA Memo Series 79, SPDO (2006)
7. Carilli, C., Rawlings, S. (eds.): Science with the Square Kilometre Array. Elsevier, Amsterdam (2004)
8. Chatterjee, R., Marscher, A.P., Jorstad, S.G., Olmstead, A.R., McHardy, I.M., Aller, M.F., Aller, H.D., et al.: Disk-jet connection in the radio galaxy 3C 120. Astrophys. J. **704**, 1689 (2009)
9. Cheung, C.C., Harris, D.E., Stawarz, Ł.: Superluminal radio features in the M87 jet and the site of flaring TeV gamma-ray emission. Astrophys. J. Lett. **663**, L65 (2007)
10. Cordes, J.M.: Survey Metrics. SKA Memo Series 109, SPDO (2009)
11. Dewdney, P., bij de Vaate, J.G., Cloete, K., Gunst, A., Hall, D., McCool, R., Roddis, N., Turner, W.: SKA Phase 1: Preliminary System Description. SKA Memo Series 130, SPDO (2010)

12. Di Matteo, T., Croft, R.A.C., Springel, V., Hernquist, L.: The cosmological evolution of metal enrichment in quasar host galaxies. Astrophys. J. **610**, 80 (2004)
13. Dokuchaev, V.I.: Joint evolution of a galactic nucleus and central massive black hole. Mon. Not. R. Astron. Soc. **251**, 564 (1991)
14. Garrett, M.A., Cordes, J.M., De Boer, D., Jonas, J.L., Rawlings, S., Schilizzi, R.T.: Concept Design for SKA Phase 1 (SKA_1). SKA Memo Series 125, SPDO (2010)
15. Haehnelt, M.G., Kauffmann, G.: Multiple supermassive black holes in galactic bulges. Mon. Not. R. Astron. Soc. **336**, L61 (2002). DOI 10.1046/j.1365-8711.2002.06056.x
16. Klöckner, H., Baan, W.A.: Understanding extragalactic hydroxyl. Astrop. Space Sci. **295**, 277 (2005)
17. Klöckner, H., Baan, W.A., Garrett, M.A.: Investigation of the obscuring circumnuclear torus in the active galaxy Mrk231. Nature **421**, 821 (2003)
18. Komissarov, S.S., Gubanov, A.G.: Relic radio galaxies: evolution of synchrotron spectrum. Astron. Astrophys. **285**, 27 (1994)
19. Lal, D.V., Lobanov, A.P., Jiménez-Monferrer, S.: Array configuration studies for the Square Kilometre Array – Implementation of figures of merit based on spatial dynamic range. SKA Memo Series 107, SPDO (2009)
20. León-Tavares, J., Lobanov, A.P., Chavushyan, V.H., Arshakian, T.G., Doroshenko, V.T., Sergeev, S.G., Efimov, Y.S., Nazarov, S.V.: Relativistic plasma as the dominant source of the optical continuum emission in the broad-line radio galaxy 3C 120. Astrophys. J. **715**, 355 (2010)
21. Lobanov, A.P.: Imaging with the SKA: Comparison to other future major instruments. SKA Memo Series 38, SPDO (2003)
22. Lobanov, A.P.: Radio spectroscopy of active galactic nuclei. MemSAIS **7**, 12 (2005)
23. Lobanov, A.P.: Binary supermassive black holes driving the nuclear activity in galaxies. MemSAI **79**, 1306 (2008)
24. Lobanov, A.P.: Physical properties of blazar jets from VLBI observations. ArXiv:1010.2856 (2010)
25. Lonsdale, C.J., Diamond, P.J., Thrall, H., Smith, H.E., Lonsdale, C.J.: VLBI images of 49 radio supernovae in Arp 220. Astrophys. J. **647**, 185 (2006)
26. Marscher, A.P., Jorstad, S.G., D'Arcangelo, F.D., Smith, P.S., Williams, G.G., Larionov, V.M., et al.: The inner jet of an active galactic nucleus as revealed by a radio-to-γ-ray outburst. Nature **452**, 966 (2008)
27. McDonald, A.R., Muxlow, T.W.B., Pedlar, A., Garrett, M.A., Wills, K.A., Garrington, S.T., Diamond, P.J., Wilkinson, P.N.: Global very long-baseline interferometry observations of compact radio sources in M82. Mon. Not. R. Astron. Soc. **322**, 100 (2001)
28. Merritt, D., Milosavljević, M.: Massive black hole binary evolution. Living Rev. Relat. **8**, 8 (2005)
29. Mezcua, M., Lobanov, A.P.: Compact radio emission in Ultraluminous X-ray sources. ArXiv:1011.0946 (2010)
30. Morganti, R., Peck, A.B., Oosterloo, T.A., van Moorsel, G., Capetti, A., Fanti, R., Parma, P., de Ruiter, H.R.: Is cold gas fuelling the radio galaxy NGC 315?Astron. Astrophys. **505**, 559 (2009)
31. Mundell, C.G., Wrobel, J.M., Pedlar, A., Gallimore, J.F.: The nuclear regions of the seyfert galaxy NGC 4151: parsec-scale H I absorption and a remarkable radio jet. Astrophys. J. **583**, 192 (2003)
32. Nipoti, C., Binney, J.: Time variability of active galactic nuclei and heating of cooling flows. Mon. Not. R. Astron. Soc. **361**, 428 (2005)
33. Peck, A.B., Taylor, G.B.: Evidence for a circumnuclear disk in 1946+708. Astrophys. J. Lett. **554**, L147 (2001)
34. Pedlar, A., Muxlow, T., Smith, R., Thrall, H., Beswick, R., Aalto, S., Booth, R., Wills, K.: OH molecules and masers in messier 82. In: Aalto, S., Huttemeister, S., Pedlar, A. (eds.) The Neutral ISM in Starburst Galaxies, Astronomical Society of the Pacific Conference Series, vol. 320, p. 183 (2004)

35. Schilizzi, R.T., Alexander, P., Cordes, J.M., Dewdney, P.E., Ekers, R.D., Gaensler, B.M., Faulkner, A.J., Hall, P.J., Jonas, J.L., Kellermann, K.I.: Preliminary Specifications for the Square Kilometre Array. SKA Memo Series 100, SPDO (2007)
36. Schinzel, F.K., Lobanov, A.P., Jorstad, S.G., Marscher, A.P., Taylor, G.B., Zensus, J.A.: Radio Flaring Activity of 3C 345 and its Connection to Gamma-ray Emission. ArXiv:1012.2820 (2010)
37. Schoenmakers, A.P., de Bruyn, A.G., Röttgering, H.J.A., van der Laan, H., Kaiser, C.R.: Radio galaxies with a 'double-double morphology' - I. analysis of the radio properties and evidence for interrupted activity in active galactic nuclei. Mon. Not. R. Astron. Soc. **315**, 371 (2000)
38. Stanghellini, C., O'Dea, C.P., Dallacasa, D., Cassaro, P., Baum, S.A., Fanti, R., Fanti, C.: Extended emission around GPS radio sources. Astron. Astrophys. **443**, 891 (2005)

Precision Astrometry: From VLBI to Gaia and SKA

Patrick Charlot

Abstract For the past 20 years, VLBI has been unique in establishing celestial reference frames based on extragalactic objects and measuring motions and distances of objects within the Galaxy with unprecedented accuracies. However, it will be challenged soon by the upcoming Gaia space astrometry mission, which will observe a billion of galactic and extragalactic objects at optical wavelengths with similar or improved astrometric accuracies compared to VLBI. In the longer term, SKA may also play a major role in this area depending on whether it includes long baselines. The paper reviews recent highlights from astrometric VLBI along with expectations from the Gaia mission within the next decade. It also draws prospects for future synergies between radio and optical astrometry in this changing context.

1 Introduction

The highest precision currently reachable in astrometry comes from interferometric observations at radio frequencies using intercontinental baselines. Through such observations, absolute positions of extragalactic radio sources are routinely obtained with sub-milliarcsecond (mas) accuracy. Over the years, continued Very Long Baseline Interferometry (VLBI) measurements have allowed the community to build the most accurate celestial frame available to date, namely the International Celestial Reference Frame (ICRF) [25]. The ICRF – and now its successor, the ICRF2 [14] – has been recognized as the fundamental IAU celestial reference frame since 1997. Additionally, the VLBI technique can also be used to measure the relative position of a target with respect to a nearby stationary calibrator source

P. Charlot (✉)
Laboratoire d'Astrophysique de Bordeaux, Université de Bordeaux, CNRS/INSU, UMR 5804, 2 rue de l'Observatoire, BP 89, 33271 Floirac Cedex, France
e-mail: charlot@obs.u-bordeaux1.fr

D. Barbosa et al. (eds.), *The Square Kilometre Array: Paving the way for the new 21st century radio astronomy paradigm*, Astrophysics and Space Science Proceedings, DOI 10.1007/978-3-642-22795-0_9, © Springer-Verlag Berlin Heidelberg 2012

with an even higher accuracy [2, 38]. By repeating measurements over the years, it is then possible to derive trigonometric parallaxes and proper motions of stars in the Milky Way with unsurpassed accuracies using this differential VLBI astrometric technique.

At optical wavelengths, there is currently no extragalactic reference frame as accurate as the ICRF or ICRF2 since astrometric accuracy from ground-based optical telescopes is poor compared to that obtained in the radio band with VLBI. The most comprehensive optical reference frame available to date is the Large Quasar Reference Frame which has positional errors peaking at 130 mas [1]. Quasar astrometry in local fields is an order of magnitude better but still far from the accuracies reached by differential VLBI astrometry [10]. The present situation, however, is about to change with the upcoming launch of the Gaia space astrometric mission which will survey the entire sky down to a magnitude limit of 20 and with targeted positional errors of few tens of microarcseconds (μas) for the brighter objects. Within the next decade, both a radio frame and an optical frame of similar accuracies will hence coexist, allowing one to develop comparative studies highly-relevant for astrophysics.

The next section presents highlights of recent VLBI astrometric results in the areas of celestial reference frames and relative astrometry and draw future prospects in the field. Section 3 reviews the current status and science goals of the Gaia mission, while Sect. 4 addresses the synergy between radio and optical astrometry.

2 VLBI Astrometry

2.1 Celestial Reference Frames

The original ICRF was based on the radio positions of 608 extragalactic sources estimated from VLBI data acquired at 8.4/2.3 GHz between 1979 and 1995 [25]. Individual ICRF source coordinates had a noise floor of 250 μas, while the axes of the frame were good to 20 μas. An additional 109 sources was added to the original frame a few years later [13]. The ICRF2, which replaced the ICRF on 1 January 2010, was generated from nearly 30 years of VLBI data accumulated during 4540 VLBI sessions conducted at 8.4/2.3 GHz between 1979 and 2009 [14]. Most of these sessions were organized in the framework of the International VLBI Service for geodesy and astrometry (IVS). A small number of such sessions also comes from dedicated surveys such as the VLBA Calibrator Survey [3, 15, 20, 34–36]. In all, 6.5 million VLBI measurements were used to construct the ICRF2. The catalog contains positions for 3,414 sources, more than five times the number of sources comprised in the ICRF (Fig. 1). The ICRF2 has a noise floor of 40 μas, some five to six times better than the ICRF, and an axis stability of 10 μas, twice as stable as the ICRF.

The VLBI extragalactic celestial reference frame is improving continuously through joint observational efforts of the VLBI community and by taking advantage

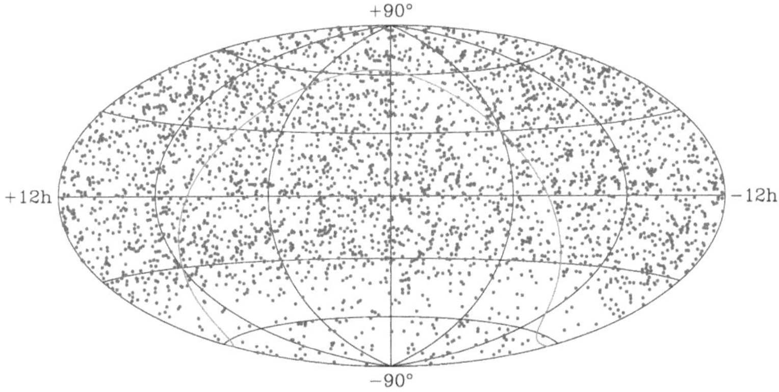

Fig. 1 Distribution of the ICRF2 sources on the celestial sphere [14]

of the latest refinements concerning modeling (e.g. troposphere) and data acquisition technology (e.g. higher recording rates). A specific effort is also targeted towards extending the frame to higher radio frequencies, chiefly at 24 and 43 GHz [12, 22] and 32 GHz [19]. Measuring source positions at these higher frequencies enables the study of frequency-dependent positional errors which may originate from extended emission farther out in the source jet and from shifts in the the radio core [11, 37]. Generally, one expects systematic errors from non-point-like source structure to be reduced at higher frequency since extended jet emission tends to fade with increasing radio frequency [12]. Current results show that the high-frequency positions agree with the standard 8.4 GHz positions at the 300 μas level, thereby indicating that the frequency-dependent positional shifts are on average below this level.

2.2 *Relative Astrometry*

As noted above, the VLBI technique may also be used to measure differential positions between two angularly-close sources by using a specific observational strategy called *phase-referencing* [2]. This strategy consists in alternating observations between a calibrator (generally an ICRF2 source) and the target source with a cycle time of a few minutes or less. The angular separation between the target and calibrator should be a few degrees at most as otherwise angular-dependent systematic errors in the directions of the target and calibrator may differ significantly and hence not fully cancel when differenced. Note that switching is not required in the case of the Japanese VERA (Very Exploration of Radio Astrometry) array which can observe two sources within 2.5° simultaneously [18]. In optimum conditions (separation calibrator-target <1°, accurate telescope locations, calibrator position and Earth's orientation), the relative position between the target and calibrator may be determined with an accuracy of a few tens of μas with this observing strategy [38].

During the past decade, major results have been derived from phase-referenced VLBI observations, among which the determination of proper motions and parallaxes of non-thermal radio-emitters in star forming regions. For example, the distance to the Orion Nebula Cluster and the Taurus region have been measured to less than 1%, including also a determination of the depth of the region for the latter [24, 26, 43, 44]. Such measurements allows one to calibrate luminosities precisely, which is mandatory to constrain stellar evolution models. Additionally, a large program has been initiated to map the Galactic structure by measuring parallaxes and proper motions of methanol masers across large portions of the Milky Way [9, 31, 40–42, 46, 47]. Ultimately, the results of this program should reinforce the foundations for the spiral structure of the Milky Way by confirming the existence of the postulated spiral arms and pinpointing their locations [40]. Another highlight was the determination of the proper motion of Sgr A*, the compact radio source at the Galactic center, with an accuracy of 0.4% [39]. Interestingly, the motion was found to be fully consistent with that expected from the motion of the Sun around the Galactic center, hence indicating that Sgr A* most probably harbors a supermassive black hole. Pushing further the technique, proper motions of the nearby galaxies M33 and IC10 have also been derived based on observations of water masers. This draws interesting prospects to study Local Group dynamics and the distribution of dark matter in those satellite galaxies of the Milky Way [7, 8].

2.3 Future Developments

The VLBI technique has been constantly improving since its inception more than 40 years ago. During the past decade, major upgrades were conducted in four areas: (1) the construction of new bigger antennas such as that in Yebes (Spain) or others in Russia and China, all in the 30–50 m diameter range, (2) the replacement of old magnetic tapes by hard disks of ever increasing capacity, (3) the increase of the recording rates up to 1 Gb/s, and (4) the development of real-time correlation by transmission of the signal from the telescopes to the correlator over optical fibre networks (e-VLBI). Such upgrades will continue in the coming years, e.g. two new antennas of the 60 m class are being built in Sardinia and in China, while recording rates of 4 Gb/s are now being considered. All these upgrades contribute to increasing the VLBI sensitivity and hence facilitating observations of weaker sources, a necessary step towards a further densification of the VLBI celestial frame.

On the geodesy side, the IVS community is moving towards a new generation system based on the use of small (12 m diameter) fast-slewing (6–12°/s) automated antennas [32]. In this new system, the entire frequency range from 2 to 14 GHz will be recorded to compensate for the small size of the antennas. For most of the ICRF sources, the required signal-to-noise ratio will be reached within 10 s. In all, the combination of short integration times and reduced source switching times (60 s at most from one side to the other side of the sky) will augment the volume of VLBI data acquired every day by one to two order of magnitudes, which in turn will result in a significant increase in precision, as inferred from simulations. While this new system was primarily designed for 1 mm geodesy and continuous monitoring of the Earth's rotation, it will also make possible re-observation of a large portion of the ICRF everyday, hence contributing to increasing its accuracy and stability. At present, about ten such next generation antennas are being built around the world, in Germany, Spain, Portugal, Australia and New Zealand, while a number of proposals for other sites are in various stages of preparation and approval [33].

3 Optical Astrometry

3.1 Overview of the Gaia Mission

The Gaia mission is an all-sky astrometric and spectro–photometric survey of celestial objects between 6th and 20th magnitude to be carried out over a period of 5–6 years. Some one billion stars, a few million galaxies, half a million quasars, and a few hundred thousand asteroids will be observed. Expected position accuracies at mean epoch are 6–10 μas for objects brighter than magnitude 14, 10–70 μas for those with magnitudes between 14 and 18, and 200 μas for those at magnitude 20 [23]. The astrometric data will be supplemented by low-resolution spectro–photometric data in the 330–1,000 nm wavelength range and, for the brighter stars, radial velocity measurements. Each object will cross one of the instrument two fields of view some 12–25 times per year at irregular intervals, providing a good temporal sampling of variability and orbital motion on all time scales from hours to years.

The procurement for the Gaia satellite (including the payload and its scientific instruments), launch and mission operations is fully funded by the European Space Agency (ESA). EADS Astrium is the prime industrial contractor for building the satellite according to the scientific and technical requirements formulated by ESA in consultation with the scientific community. The launch is scheduled for the beginning of 2013. Transfer of the satellite to its orbit around the Sun–Earth L_2 Lagrange point (1.5 million km from the Earth) and commissioning will take up 6 months, after which routine science operations will start. This operational phase will last 5 years with a possible 1 year extension. The Data Processing and Analysis Consortium, formed in 2006, is in charge of designing, implementing and running the required software system for processing the Gaia data [29]. Currently, there are about 400 people involved in the Gaia project, originating from 22 countries, not counting the industrial contractors. The final Gaia catalog is expected a few years after the observation ends but with intermediate results released during the operational phase.

3.2 Gaia Science Goals

The scientific objectives of the Gaia mission cover an extremely wide range of topics in galactic and stellar astrophysics, solar system and exoplanet science, as well as in the areas of reference frames and fundamental physics [23].

A primary goal of the mission is to study the structure and evolution of the Galaxy by correlating the spatial distribution and kinematics of stars with their astrophysical properties. The determination of number densities and three-dimensional motions for large, volume-complete samples of stars will allow one to trace the galactic potential and hence the distribution of matter (including dark matter) in unprecedented details. Additionally, the combination of luminosity and colour information will constrain the history of star formation, which in combination with kinematic data may provide new insights into how the Galaxy was assembled and evolved.

Gaia will have an equally large impact in stellar astrophysics. By measuring distances to millions of stars to <1 % accuracy, it will boost all studies that depend on the intrinsic properties of stars. In particular, it will provides stringent tests of stellar structure models, driving further the development of improved theoretical models of stellar atmospheres, interiors and evolution. Taking advantage of the highly-accurate astrometric data, planetary companions may be detected for millions of nearby stars, thereby providing reliable statistics for the occurrence of exoplanetary systems and their characteristics (masses and orbital elements of the detected companions).

A huge contribution will also be made to Solar System physics through the systematic observations of 350,000 asteroids brighter than magnitude 20. While only 50,000 new objects are expected, the large number of observations (some 60 epochs per object) and the high accuracy of these observations will permit the determination of extremely accurate osculating elements and hence the investigation of orbit families, their dynamical evolution, and the masses of the individual asteroids.

Taking advantage of the 500,000 quasars brighter than magnitude 20 that will be detected, Gaia will establish for the first time a very accurate, dense and faint optical reference frame, directly at optical wavelengths. Simulations show that the residual spin of the Gaia frame should be determined to 0.5 μas/year, assuming a "clean sample" of 10,000 defining sources [27]. In addition, a number of General Relativity tests will be possible, including determination of the PPN parameters β and γ and solar quadrupole parameter J_2, verification of the Local Lorentz invariance, potential variations of the gravitational constant and deviations from Newtonian gravitational law, or the existence of low-frequency gravitational waves [16, 17, 30].

4 Synergy Between Radio and Optical Astrometry

4.1 Aligning the VLBI and the Gaia Frame

The future realization of a highly-accurate extragalactic reference frame at optical wavelengths by the Gaia space astrometric mission raises the issue of the alignment of this frame with the ICRF2 (or its successor by the time the Gaia frame is built). Such alignment, to be obtained with the highest accuracy, requires a large number of sources common to the two frames, i.e. radio-loud objects whose positions are determined accurately both by Gaia and from VLBI. As noted in [28], this implies that the link sources must be brighter than magnitude 18 (for the highest Gaia position accuracy). Additionally, they should have compact VLBI structures on milliarcsecond scales for the highest VLBI position accuracy. Based on these considerations, a study revealed that only 10% of the ICRF (70 sources) meet these criteria [4], which prompted the development of a new specific VLBI observing program dedicated to finding and characterizing additional such suitable sources.

For this purpose, a sample of 447 optically-bright (magnitude ≤ 18) candidate radio sources, most of which located in the northern sky, was selected from a deeper radio survey and observed with VLBI [5]. The observing strategy for identifying the suitable optical-radio link sources comprises three successive steps dedicated to (1) assess the VLBI detectability of the targets, (2) image them, and (3) measure their astrometric positions. Results for step 1 indicate a nearly 90% detection rate, with 398 sources detected out of the 447 targets comprised in the original sample [5]. Imaging 105 of these sources (step 2) reveals that about half of them qualify for the alignment in terms of source compactness [6]. See Fig. 2 for a sample of 8 GHz images for six of the targets. Assuming similar statistics, it is anticipated that about 200 suitable sources should be identified from this survey once the project

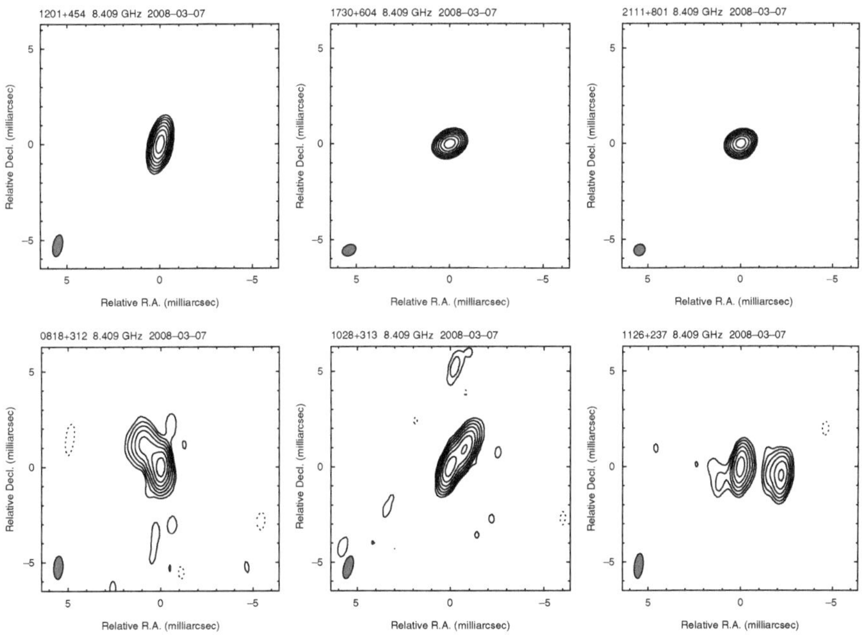

Fig. 2 8 GHz VLBI images of candidate radio sources for the alignment with the future Gaia optical frame [6]. *Upper panel:* Three sources suitable for this alignment (point-like sources). *Lower panel:* Three sources not suitable for this alignment (sources with extended VLBI structures)

is completed. Similar work should also be undertaken for the southern hemisphere in order to ensure a complete sky coverage of the sources serving in the alignment.

4.2 Prospects for Astrophysics

An issue of potential trouble – but of interest – when aligning the VLBI and Gaia frames is whether the optical emission and radio emission are superimposed in quasars. Recent estimates of such optical-radio "core shifts" indicate they amount to 100 μas on average [21], which is significant when considering the few-tens-of-μas accuracy of the Gaia and VLBI frames. While potentially impacting the alignment between the frames, differences between the optical and radio positions may also offer a unique opportunity to directly determine these core-shifts and probe the geometry of quasars in the framework of unified theories of active galactic nuclei [45]. Such measurements may be able to locate the optical region relative to the relativistic radio jet and determine whether the dominant optical emission originates from the accretion disk or the inner portion of the jet. In this respect, the construction of reference frames at higher radio frequencies [19, 22] may also bring important clues in that they provide intermediate positions between the ICRF and Gaia positions.

4.3 Potential Role of SKA

As noted above, the current VLBI frame comprises a few thousands sources, which is less than 1% of the anticipated number of extragalactic objects that Gaia will detect. Therefore, it is clear that VLBI will not be able to compete with Gaia in terms of source density – even though its current astrometric accuracy is now approaching that of Gaia – as it would require sensitivities and resources that are out of reach of existing VLBI networks. On the other hand, the Square Kilometre Array (SKA) with its nanoJy extreme sensitivity and survey capability might be able to scan the sky in the same way as Gaia and hence build the radio counterpart of the Gaia optical frame when it comes into operations beyond 2020. To this end, SKA must have long baselines ($\geq 5,000$ km) and be able to reach an observing frequency of 10 GHz at least, as otherwise the astrometric accuracy will not be high enough. With such prospects in mind, the VLBI community should start developing innovative observing schemes to use SKA for astrometric applications. For example, one can think of observing several targets simultaneously with different sub-arrays to reduce tropospheric errors or using a small subset of the SKA to monitor the Earth's rotation permanently without impacting the science conducted at the same time with the rest of the array, but other ideas can be put forward as well. It is also to be noted that in-beam calibrators for phase-referencing will always be available with SKA.

5 Conclusion

While VLBI has been the only player in the area of precision astrometry, the coming decade will bring new or upgraded instrumental facilities that will have a strong impact on astrometrical research. In particular, the launch of the Gaia space mission in 2013 will revolutionize optical astrometry. On the radio side, upgrades of the current VLBI networks towards higher sensitivities and automation are continuing at a high pace while SKA may take over in the longer term. Thanks to the Gaia satellite, positions of celestial objects whether galactic or extragalactic will be measured for the first time with a few tens of μas accuracy at both optical and radio wavelengths. Comparison of these positions will probe directly the geometry and physics of the targets as never done before, which holds promises for new discoveries.

Acknowledgements I would like to thank François Mignard, Géraldine Bourda, Sébastien Lambert and Chris Jacobs for providing material for this presentation. The section on Gaia also made use of material from Lennart Lindegren.

References

1. Andrei, A.H., Souchay, J., Zacharias, N., Smart, R.L., Viera Martins, R., da Silva Neto, D.N., Camargo, J.I.B., Assafin, N., Barache, C., Bouquillon, S., Penna, J.L., Taris, F.: The large quasar reference frame (LQRF). An optical representation of the ICRS. Astron. Astrophys. **505**, 385–404 (2009)
2. Beasley, A.J., Conway, J.E.: VLBI phase-referencing. In: Zensus, J.A., Diamond, P.J., Napier P.J. (eds.) Very long baseline interferometry and the VLBA. Astron. Soc. Pac. Conf. Ser. **82**, 328–343 (1995)
3. Beasley, A.J., Gordon, D., Peck, A.B., Petrov, L., McMillan, D.S., Fomalont, E.B., Ma, C.: The VLBA calibrator survey-VCS1. Astrophys. J. Suppl. **141**, 13–21 (2002)
4. Bourda, G., Charlot, P., Le Campion, J.-F.: Astrometric suitability of optically-bright ICRF sources for the alignment with the future gaia celestial reference frame. Astron. Astrophys. **490**, 403–408 (2008)
5. Bourda, G., Charlot, P., Porcas, R. Garrington, S.T.: VLBI observations of optically-bright extragalactic radio sources for the alignment of the radio frame with the future gaia frame. I. Source Detection. Astron. Astrophys. **520**, A113 (2010)
6. Bourda, G., Collioud, A., Charlot, P., Porcas, R., Garrington, S.T.: VLBI observations of optically-bright extragalactic radio sources for the alignment of the radio frame with the future gaia frame. II. Imaging Candidate Sources. Astron. Astrophys. **526**, A102 (2011)
7. Brunthaler, A., Reid, M.J., Falcke, H., Greenhill, L.J., Henkel, C.: The geometric distance and proper motion of the triangulum galaxy (M33). Science **307**, 1440–1443 (2005)
8. Brunthaler, A., Reid, M.J., Falcke, H., Henkel, C., Menten, K.M.: The proper motion of the local group galaxy IC 10. Astron. Astrophys. **462**, 101–106 (2007)
9. Brunthaler, A., Reid, M.J., Menten, K.M., Zheng, X.W., Moscadelli, L., Xu, Y.: Trigonometric parallaxes of massive star-forming regions: V. G23.01−0.41 and G23.44−0.18. Astrophys. J. **693**, 424–429 (2009)
10. Camargo, J.I.B., Daigne, G., Ducourant, C., Charlot, P.: Near-infrared astrometry and photometry of southern ICRF quasars. Astron. Astrophys. **435**, 1135–1146 (2005)

11. Charlot, P.: Radio source structure in astrometric and geodetic very long baseline interferometry. Astron. J. **99**, 1309–1326 (1990)
12. Charlot, P., Boboltz, D.A., Fey, A.L., Fomalont, E.B., Geldzhaler, B.J., Gordon, D., Jacobs, C.S., Lanyi, G.E., Ma, C., Naudet, C.J., Romney, J.D., Sovers, O.J., Zhang, L.D.: The celestial reference frame at 24 and 43 GHz. II. Imaging. Astron. J. **139**, 1713–1170 (2010)
13. Fey, A.L., Ma, C., Arias, E.F., Charlot, P., Feissel-Vernier, M., Gontier, A.-M., Jacobs, C.S., Li, J., McMillan, D.S.: The second extension of the international celestial reference frame: ICRF-Ext. 2. Astron. J. **127**, 3587–3608 (2004)
14. Fey, A.L., Gordon D., Jacobs, C.S. (eds.): The Second Realization of the International Celestial Reference Frame by Very Long Baseline Interferometry. IERS Technical Note 35, Verlag des Bundesamts für Kartographie und Geodäsie, Frankfurt am Main (2010)
15. Fomalont, E.B., Petrov, L., McMillan, D.S., Gordon, D., Ma, C.: The second VLBA calibrator survey: VCS2. Astron. J. **126**, 2562–2566 (2003)
16. Hestroffer, D., Mouret, S., Mignard, F., Tanga, P., Berthier, J.: Gaia and the asteroids: local test of GR. In: Klioner, S.A., Seidelman, P.K., Soffel M.H. (eds.) Relativity in Fundamental Astronomy, Proceedings of the IAU Symposium No. 261, 325–330 (2009)
17. Hobbs, D., Holl, B., Lindegren, L., Raison, F., Klioner, S., Butkevich, A.: Determining PPN γ with Gaia's astrometric core solution. In: Klioner, S.A., Seidelman, P.K., Soffel, M.H. (eds.) Relativity in Fundamental Astronomy, Proceedings of the IAU Symposium No. 261, 315–319 (2009)
18. Honma, M., Fujii, T., Hirota, T. et al.: First fringe detection with VERA's dual-beam system and its phase-referencing capability. Publ. Astron. Soc. Japan **55**, L57–L60 (2003)
19. Jacobs, C.S., Sovers, O.J., Clark, J.E., Garcia-Miro, C., Horiuchi, S., Moll, V.E., Skjerve, L.J.: X/Ka celestial frame. Astron. J. in preparation (2010)
20. Kovalev, Y.Y., Petrov, L., Fomalont, E.B., Gordon, D.: The fifth VLBA calibrator survey - VCS5. Astron. J. **133**, 1236–1242 (2007)
21. Kovalev, Y.Y., Lobanov, A.P., Pushkarev, A.B., Zensus, J.A.: Opacity in compact extragalactic radio sources and its effect on astrophysical and astrometric studies. Astron. Astrophys. **483**, 759–768 (2008)
22. Lanyi, G.E., Boboltz, D.A., Charlot, P., Fey, A.L., Fomalont, E.B., Geldzhaler, B.J., Gordon, D., Jacobs, C.S., Ma, C., Naudet, C.J., Romney, J.D., Sovers, O.J., Zhang, L.D.: The celestial reference frame at 24 and 43 GHz. I. Astrometry. Astron. J. **139**, 1695–1712 (2010)
23. Lindegren, L.: Gaia: astrometric performance and current status of the project. In: Klioner, S.A., Seidelman, P.K., Soffel M.H. (eds.) Relativity in Fundamental Astronomy, Proceedings of the IAU Symposium No. 261, 296–305 (2009)
24. Loinard, L., Torres, R.M., Mioduszewski, A.J., Rodríguez, L.F., González-Lópezlira, R.A., Lachaume, R., Vázquez, V., González, E.: VLBA determination of the distance to nearby star-forming regions. I. The distance to T Tauri with 0.4% accuracy. Astrophys. J. **671**, 546–554 (2007)
25. Ma, C., Arias, E.F., Eubanks, T.M., Fey, A.L., Gontier, A.-M., Jacobs, C.S., Sovers, O.J., Archinal, B.A., Charlot, P.: The international celestial reference frame as realized by very long baseline interferometry. Astron. J. **116**, 516–546 (1998)
26. Menten, K.M., Reid, M.J., Forbrich, J., Brunthaler, A.: The distance of the orion nebula. Astron. Astrophys. **474**, 515–520 (2007)
27. Mignard, F.: Observations of QSOs and reference frame with gaia. In: Bienaymé O., Turon, C. (eds.) Proceedings of the Conference: Gaia: A European Space project, EAS Publ. Ser. **2**, 327–339 (2002)
28. Mignard, F.: Realization of the inertial frame with gaia. In: Gaume, R., McCarthy, D.D., Souchay, J. (eds.) Proceedings of the Joint Discussion 16, The International Celestial Reference System, Maintenance and Future Realization, IAU XXV General Assembly, USNO, Washington, pp. 133–140 (2003)
29. Mignard, F., Bailer-Jones, C., Bastian, U., Drimmel, R., Eyer, L., Katz, D., van Leeuwen, F., Luri, X., O'Mullane, W., Passot, X., Pourbaix, D., Prusti, T.: Gaia: organisation and challenges for the data Processing. In: Jin, W.J., Platais, I., Perryman, M.A.C. (eds.) A Giant Step: from

Milli- to Micro-Arcsecond Astrometry, Proceedings of the IAU Symposium No. 248, 224–230 (2008)
30. Mignard, F., Klioner, S.S.: Gaia: relativistic modelling and testing. In: Klioner, S.A., Seidelman, P.K., Soffel, M.H. (eds.) Relativity in Fundamental Astronomy, Proceedings of the IAU Symposium No. 261, 306–314 (2009)
31. Moscadelli, L., Reid, M.J., Menten, K.M., Brunthaler, A., Zheng, X.W., Xu, Y.: Trigonometric parallaxes of massive star-forming regions: II. CEP A and NGC 7538. Astrophys. J. **693**, 406–412 (2009)
32. Petrachenko, W., Niell, A.E., Behrend, D., et al.: Design Aspects of the VLBI2010 System – Progress Report of the IVS VLBI2010 Committee. NASA/TM-2009-214180 (2009)
33. Petrachenko, W.: VLBI2010: An overview. In: Behrend D., Baver K.D. (eds.) IVS 2010 General Meeting Proceedings – VLBI2010: From Vision to Reality, NASA/CP-2010-215864, 3–7 (2010)
34. Petrov, L., Kovalev, Y.Y., Fomalont, E.B., Gordon, D.: The third VLBA calibrator survey: VCS3. Astron. J. **129**, 1163–1170 (2005)
35. Petrov, L., Kovalev, Y.Y., Fomalont, E.B., Gordon, D.: The fourth VLBA calibrator survey: VCS4. Astron. J. **131**, 1872–1879 (2006)
36. Petrov, L., Kovalev, Y.Y., Fomalont, E.B., Gordon, D.: The sixth VLBA calibrator survey: VCS6. Astron. J. **136**, 580–585 (2008)
37. Porcas, R.W.: Radio astrometry with chromatic AGN core positions. Astron. Astrophys. **505**, L1–L4 (2009)
38. Pradel, N., Charlot, P., Lestrade, J.-F.: Astrometric accuracy of phase-referenced observations with the VLBA and EVN. Astron. Astrophys. **452**, 1099–1106 (2006)
39. Reid, M.J., Brunthaler, A.: The proper motion of sagittarius A*. II. The mass of sagittarius A*. Astrophys. J. **616**, 872–884 (2004)
40. Reid, M.J., Menten, K.M., Brunthaler, A., Zheng, X.W., Moscadelli, L., Xu, Y.: Trigonometric parallaxes of massive star-forming regions: I. S252 & G232.6+1.0. Astrophys. J. **693**, 397–405 (2009)
41. Reid, M.J., Menten, K.M., Zheng, X.W., Brunthaler, A., Moscadelli, L., Xu, Y., Zhang, B., Sato, M., Honma, M., Hirota, T., Hachisuka, K., Choi, Y.K., Moellenbrock, G.A., Bartkiewicz, A.: Trigonometric parallaxes of massive star-forming regions: VI. Galactic structure, fundamental parameters, and noncircular motions. Astrophys. J. **700**, 137–148 (2009)
42. Sanna, A., Reid, M.J., Moscadelli, L., Dame, T.M., Menten, K.M., Brunthaler, A., Zheng, X.W., Xu, Y.: Trigonometric parallaxes of massive star-forming regions: VII. G9.62+0.20 and the expanding 3 kpc Arm. Astrophys. J. **706**, 464–470 (2009)
43. Torres, R.M., Loinard, L., Mioduszewski, A.J., Rodríguez, L.F.: VLBA determination of the distance to nearby star-forming regions. II. Hubble 4 and HDE 283572 in Taurus. Astrophys. J. **671**, 1813–1819 (2007)
44. Torres, R.M., Loinard, L., Mioduszewski, A.J., Rodríguez, L.F.: VLBA determination of the distance to nearby star-forming regions. III. HP Tau/G2 and the three-dimensional structure of Taurus. Astrophys. J. **698**, 242–249 (2009)
45. Urry, C.M., Padovani, P.: Unified schemes for radio-loud active galactic nuclei. Publ. Astron. Soc. Pac. **107**, 803–845 (1995)
46. Xu, Y., Reid, M.J., Menten, K.M., Brunthaler, A., Zheng, X.W., Moscadelli, L.: Trigonometric parallaxes of massive star-forming regions: III. G59.7+0.1 and W 51 IRS2. Astrophys. J. **693**, 413–418 (2009)
47. Zhang, B., Zheng, X.W., Reid, M.J., Menten, K.M., Xu, Y., Brunthaler, A., Moscadelli, L., Brunthaler, A.: Trigonometric parallaxes of massive star-forming regions: IV. G35.20−0.74 and G35.20−1.74. Astrophys. J. **693**, 419–423 (2009)

Ultra Steep Spectrum Radio Sources in the Lockman Hole: SERVS Identifications and Redshift Distribution at the Faintest Radio Fluxes

L. Bizzocchi, J. Afonso, E. Ibar, M. Grossi, C. Simpson, S. Chapman, M.J. Jarvis, H. Rottgering, R.P. Norris, J. Dunlop, R.J. Ivison, H. Messias, J. Pforr, M. Vaccari, N. Seymour, P. Best, E. Gonz, D. Farrah, J.-S. Huang, M. Lacy, C. Marastron, L. Marchetti, J.-C. Mauduit, S. Oliver, D. Rigopoulou, S.A. Stanford J. Surace, and G. Zeimann

Abstract Ultra Steep Spectrum (USS) radio sources have been successfully used to select powerful radio sources at high redshifts ($z > 2$). Typically restricted to large-sky surveys and relatively bright radio flux densities, it has gradually become possible to extend the USS search to sub-mJy levels, thanks to the recent appearance of sensitive low-frequency radio facilities. Here we present a first detailed analysis of the nature of the faintest USS sources. By using GMRT and VLA radio observations of the Lockman Hole (LH) at 610 MHz and 1.4 GHz, a sample of 58 micro-Jansky USS sources is assembled. Deep infrared data at 3.6 and 4.5 μm from the *Spitzer Extragalactic Representative Volume Survey* (SERVS) is used to reliably identify counterparts for 48 (83%) of these sources, showing an average magnitude of [3.6] = 19.7 mag(AB). Spectroscopic redshifts for 14 USS sources, together with photometric redshift estimates, improved by the use of the deep SERVS data, for a further 19 objects, show redshifts ranging from $z = 0.1$ to $z = 2.8$, peaking at $z \sim 0.6$ and tailing off at high redshifts.

1 Radio Sample and USS Selection

The USS sources were selected from the Lockam Hole radio surveys [1] at 610 MHz (15 μJy beam^{-1} rms) and 1.4 GHz (6 μJy beam^{-1} rms). The robustness of the sample were strenghtened by limiting the 610 MHz radio sample to $S_{610\,\mathrm{MHz}} \geq 100\,\mu$Jy. Cross-correlation gave spectral indices, α_{610}^{1400} ($S_\nu \propto \nu^\alpha$) for 662 sources of the 610 MHz-catalogue, with an upper limit obtained for further 193 sources (Fig. 1). The USS source sample was then selected by imposing a conservative threshold of $\alpha_{610}^{1400} \leq -1.3$, which yielded 58 objects. The sample will not be

L. Bizzocchi (✉)
CAAUL, Observatório Astronómico de Lisboa, Tapada da Ajuda, 1349-018, Lisboa, Portugal
e-mail: bizzocchi@oal.ul.pt

D. Barbosa et al. (eds.), *The Square Kilometre Array: Paving the way for the new 21st century radio astronomy paradigm*, Astrophysics and Space Science Proceedings, DOI 10.1007/978-3-642-22795-0_10, © Springer-Verlag Berlin Heidelberg 2012

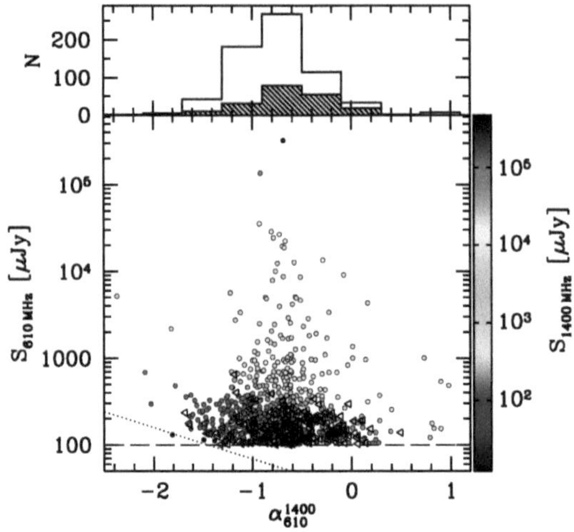

Fig. 1 Flux density at 610 MHz vs. radio spectral index for faint radio sources in the Lockman Hole. *Open symbols* are upper limits in α, representing sources with no 1.4 GHz detection. The *dashed line* represents the adopted (robust) lower limit for $S_{610\,\mathrm{MHz}}$, while the *dotted line* shows the locus of points with $S_{1400\,\mathrm{MHz}} = 30\,\mu$Jy, corresponding to the [1] 5σ peak flux limit at this frequency. The histogram at the top is the distribution of α, with the hashed column representing upper limits

complete, given the α measurement errors, but it will be representative of the sub-mJy USS population, allowing for a first comprehensive study of its nature.

2 SERVS IR Counterparts and z Distribution

For the source identification we use the LH infrared data from SERVS [2], which provides deep and uniform coverage of the entire region at 3.6 and 4.5 μm, adopting the likelihood ratio method of Sutherland and Saunders [3]. For each USS source, the 3.6 μm band identification with the highest reliability above 75% was taken as the real counterpart. 48 out of the 58 (83%) USS sources have an IRAC identification, the median AB magnitude being [3.6] = 19.7 mag (Fig. 2, left panel). USS sources unidentified in SERVS are shown as filled circles in the histogram above at their magnitude 5σ lower limit. The shaded histogram represents sources with no available redshift estimate. Among the 48 USS sources with IRAC detection, we have found 14 spectroscopic redshift determinations from various past or ongoing spectroscopic surveys in the LH, whereas for 19 further sources a photometric redshift estimate is possible from 5 or more photometric measurements in the 0.36–4.5 μm range, using data from WFC/INT, UKIDSS/DXS, SERVS and SWIRE, merged into a SERVS-selected multi-wavelength catalog (SERVS Data Fusion [4]). The redshift distribution of USS sources is presented in the right panel of Fig. 2: it shows a significant presence of high redshift sources, with a peak at

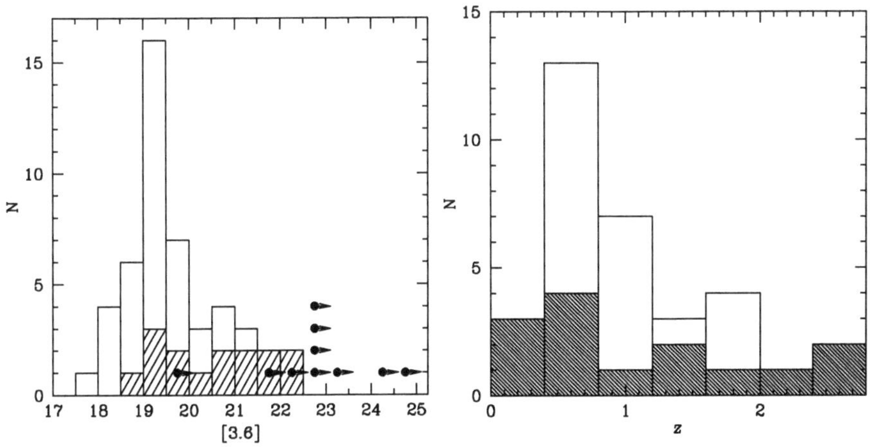

Fig. 2 *Left*: [3.6]-band AB magnitude distribution for USS sources in the Lockman Hole. Sources with no available redshift estimate are indicated by the filled histogram. The *filled circles* and *attached arrows* indicate the USS sources with no IRAC detection, placed at the bin location of their 5 σ [3.6] magnitude lower limit. *Right*: Redshift distribution for radio-faint USS sources in the Lockman Hole. *Filled histogram* denotes sources with a spectroscopic redshift determination, while the *open histogram* refers to photometric redshift estimates

$z \sim 0.6$ and extending beyond $z \sim 2$. The sources without a z estimate mostly correspond to the faintest [3.6] magnitudes and are potentially the highest redshift sources in our sample.

3 SKADS Simulated Sky

Using the SKADS Simulated Skies (S^3) simulations [5], we looked at the prediction for a radio survey reaching a detection sensitivity of 100 μJy at 610 MHz over 0.6 deg^2, which will be directly comparable with the current work. Countrariwise to higher flux densities (where powerful AGN are dominant), in the sub-mJy regime SF galaxies along with lower luminosity AGN dominate the radio sky, with FRI and RQ sources contributing significantly at $z < 2$. At higher redshifts ($z > 2$), FR Is dominate the survey model detections. The redshift distribution of our USS sources in the current work is remarkably similar to the FR I and RQ redshift distribution from the S^3 simulations peaking just below $z \sim 1$ and tailing-off above this redshift. The ~40 per cent USS sources in the Lockman Hole sample with no redshift estimates, if indeed at higher redshifts, would match well the model predictions, which place several tens of FRI sources at $z > 2$ for this survey. Interestingly, the efficiency of the USS technique at sub-mJy flux densities for the selection of very high redshift galaxies may still be significantly high – possibly even higher than that for radio surveys at much higher flux densities. Follow up spectroscopic observations of these sources are now ongoing (Fig. 3).

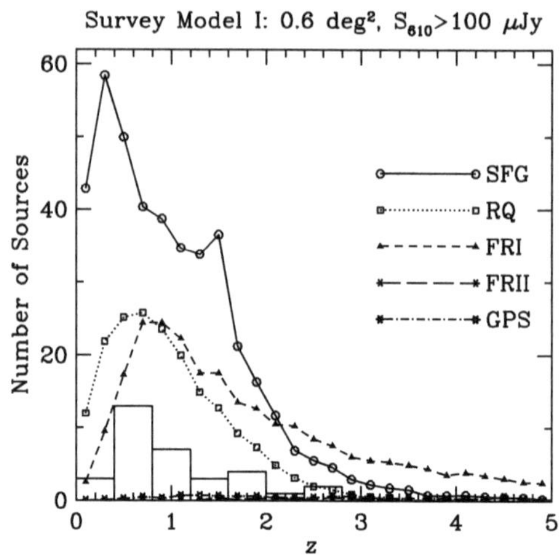

Fig. 3 Predictions from the SKADS Simulated Skies models for the redshift distributions of radio source populations for a radio survey reaching a detection sensitivity of 100 μJy at 610 MHz over 0.6 deg^2, similar to the Lockman Hole radio survey considered in the present work. The plot, also displays the observed redshift distribution for USS sources (*open histogram*)

References

1. Ibar, E., Ivison, R.J., Biggs, A.D., et al.: Mon. Not. R. Astron. Soc. 397, 281 (2009)
2. Lacy, M., Farrah, D., Mauduit, J.-C., et al.: Astrophys. J. submitted (2010)
3. Sutherland, W., Saunders, W.: Mon. Not. R. Astron. Soc. 259, 413 (1992)
4. Vaccari, M., et al.: in preparation (2010)
5. Wilman, R.J., Miller, L., Jarvis, M.J., et al.: Mon. Not. R. Astron. Soc. 388, 1335 (2008)

Probing the Very First Galaxies with the SKA

M.B. Silva, M.G. Santos, J.R. Pritchard, R. Cen, and A. Cooray

Abstract We describe the use of the high redshift 21 cm signal to probe the very first galaxies to appear in the Universe. Using fast large volume simulations of the pre-Reionization epoch we have shown that the Lyman alpha radiation emitted from these young galaxies makes a strong contribution to the 21 cm signal on large scales at $z \sim 20$. With the current setup the Square Kilometre Array (SKA) should be able to measure this signal, therefore making it probably the only telescope capable of giving us detailed information about the radiation emitted by the first stars and characterizing their host galaxies. SKA-pathfinders with ~10% of the full collecting area should be capable of making a statistical detection of the 21 cm power spectrum at redshifts $z < 20$. We then discuss the use of the redshift space distortions as a way to further constrain the Lyman alpha signal and demonstrate that they can be used as a model independent way to extract this signature with the SKA.

1 Introduction

The period of formation of the very first stars is one of the least understood epochs in the history of the Universe. The 21 cm spin-flip transition of neutral hydrogen at high redshifts has the potential to open a new observational window to study this early period, where the most distant, first galaxies reside, and even beyond that.

Although the high redshift-end of the reionization process will be inaccessible to the first generation experiments, such as the Low Frequency Array (LOFAR[1])

[1] http://www.lofar.org.

M.B. Silva (✉)
CENTRA, Departamento de Física, Instituto Superior Técnico, 1049-001 Lisboa, Portugal
e-mail: martasilva85@gmail.com

D. Barbosa et al. (eds.), *The Square Kilometre Array: Paving the way for the new 21st century radio astronomy paradigm*, Astrophysics and Space Science Proceedings, DOI 10.1007/978-3-642-22795-0_11, © Springer-Verlag Berlin Heidelberg 2012

and the Murchison Widefield Array (MWA[2]), the second generation experiments such as the Square Kilometre Array (SKA[3]) should have enough collecting area to statistically probe the signal.

With a wide frequency coverage, observations of the 21 cm signal will provide a 3-dimensional tomographic view of the inter-galactic medium (IGM), before and during the cosmological reionization [9]. As soon as the first galaxies appear, at a redshift of $z \sim 20 - 30$ in the standard cold dark matter cosmological model [13], photons emitted at frequencies between the Lyα and Lyman limit quickly couple the spin temperature of the hyperfine level population to the IGM gas temperature through the Wouthuysen-Field effect [8, 34]. Provided the IGM has been cooling adiabatically (X-ray heating is still negligible during this early stage of galaxy formation), the 21 cm signal will be strong and observed in absorption against the cosmic microwave background (CMB). Because the Lyα coupling depends on the local Lyα radiation, which is correlated with galaxies, the 21 cm signal fluctuations will trace the location, mass, emission spectra, luminosity and redshift evolution of high redshift galaxies.

On the theoretical front, analytical models have been useful in predicting the probable evolution of the high redshift signal [4, 22, 23], but rely on linear approximations and require further comparison with simulations. Numerical simulations, on the other hand, can provide a self-consistent treatment of the Lyα radiative transfer, taking into account effects such as the scattering in the Lyα line wings [1, 30], although they are typically slow to run and are limited to small volumes ($\lesssim$100 Mpc/h). Recent developments using semi-numerical algorithms have allowed the rapid generation of the high redshift 21 cm signal during the pre-reionization epoch [19, 24, 26] in large boxes, while maintaining the 3-d structure of the signal on small scales as seen in the full numerical simulations.

In this paper we will generate fast simulations as described in [26] to analyze the dependence of the overall 21 cm signal on several parameters related to the first galaxies in the Universe and consider the ability of an experiment like SKA to constrain these parameters. The layout of this paper is as follows. We start in Sect. 2 by describing the 21 cm signal and the free parameters of the model that affect the Lyα fluctuations. In Sect. 3 we present the experimental setup used for SKA, calculating the expected error on the 3-d power spectrum. In Sect. 4 we study the possibility of constraining the signal in a model independent way, using the redshift-space distortions and the corresponding measurements on the 3-d power spectrum as a function of the angle with respect to the line of sight. We end with our conclusions and a discussion of the prospects for SKA in Sect. 5.

Throughout this paper where cosmological parameters are required we use the standard set of values $\Omega_m = 0.28$, $\Omega_\Lambda = 0.72$, $\Omega_b = 0.046$, $H = 100h$ km s^{-1}Mpc^{-1} (with $h = 0.7$), $n_S = 0.96$, and $\sigma_8 = 0.82$ [13].

---

[2] http://www.mwatelescope.org.

[3] http://www.skatelescope.org.

2 The 21 cm Signal: Lyα Fluctuations

The 21 cm brightness temperature corresponds to the change in the intensity of the CMB radiation due to absorption or emission when it travels through a patch of neutral hydrogen. It is given, at an observed frequency ν in the direction $\hat{\mathbf{n}}$, by (see e.g. [24])

$$\delta T_b(\nu) \approx 23 x_{HI}(1+\delta)\left(1-\frac{T_\gamma}{T_S}\right)\left(\frac{h}{0.7}\right)^{-1}\left(\frac{\Omega_b h^2}{0.02}\right)\times$$
$$\left[\left(\frac{0.15}{\Omega_m h^2}\right)\left(\frac{1+z}{10}\right)\right]^{1/2}\left(\frac{1}{1+1/H\, dv_r/dr}\right)\text{mK}, \quad (1)$$

where $x_{\rm HI}$ is the fraction of neutral hydrogen (mass weighted), dv_r/dr is the comoving gradient of the line of sight component of the comoving velocity and we use δ_a for the fractional value of the quantity a ($\delta_a \equiv \frac{a-\langle a\rangle}{\langle a\rangle}$) with δ for the fluctuation in the matter density.

The spin temperature (T_S) is coupled to the hydrogen gas temperature (T_K) through the spin-flip transition, which can be excited by collisions or by the absorption of Lyα photons (Wouthuysen-Field effect) and we can write:

$$1-\frac{T_\gamma}{T_S}=\frac{x_{tot}}{1+x_{tot}}\left(1-\frac{T_\gamma}{T_K}\right), \quad (2)$$

where $x_{tot} = x_\alpha + x_c$ is the sum of the radiative and collisional coupling parameters and we are already assuming that the color temperature of the Lyα radiation field at the Lyα frequency is equal to T_K. Note that although we talk here about Lyα radiation, in effect we consider the contribution from all the photons up to the Lyman limit frequency, since photons redshifting into Lyman series resonances can produce Lyα photons as a result of atomic cascades [11,21].

When the coupling to the gas temperature is negligible (e.g. $x_{tot} \sim 0$), $T_S \sim T_\gamma$ and there is no signal. On the other hand, for large x_{tot}, T_S simply follows T_K (Fig. 1). Figure 1 also shows the evolution of the gas temperature with redshift, where we can see that most of the IGM is heated above 100 K for $z \leq 11$ and the evolution of the average brightness temperature with redshift, which is observed in absorption down to $z = 12$ where it becomes approximately zero when $T_K \approx T_{CMB}$.

2.1 Simulations

We use the code SimFast21[4] described in [26] to calculate the spin temperature and the corresponding 21 cm signal. The code starts by calculating the linear density

[4]Available at http://simfast21.org.

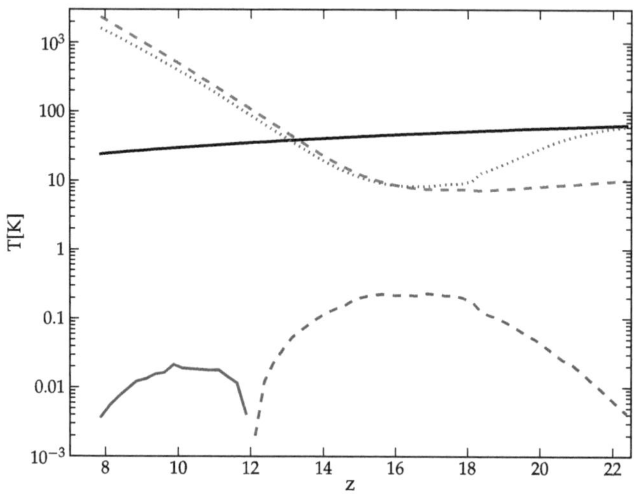

Fig. 1 *Left*: The evolution of the gas temperature, spin temperature and brightness temperature with redshift. *Solid black line* – CMB; *Red dashed* – gas temperature; *Blue dotted* – spin temperature; *Dark blue solid/dashed* (below) – brightness temperature

Table 1 Simulations used in the analysis

	Size	Resolution (halos)	Resolution (Lyα)
S1	1,000 Mpc	0.56 Mpc	1.667 Mpc
S2	143 Mpc	0.09 Mpc	0.186 Mpc
S3 (N-body)	143 Mpc	–	0.186 Mpc

field in a box followed by the corresponding halo and velocity fields. Then, it computes the nonlinear density and halo fields, the collapsed mass distribution, the corresponding star formation rate, the IGM gas temperature, the Lyα coupling, the collisional coupling and finally the 21 cm brightness temperature accounting for all the contributions including the correction due to redshift space distortions using the nonlinear velocity field. In order to probe the full range of k space, we generated two end to end simulations using the SimFast21 code, one large simulation with 1Gpc (**S1**) and another with higher resolution but smaller 143 Mpc in size (simulation **S2**). For comparison, we also applied our Lyα calculation to the output (star formation rate, matter density) from a numerical simulation [33], thus providing a check of the approximation used to generate the halo mass function. We call this simulation of the 21 cm signal, **S3**, with a size of 143 Mpc. Table 1 shows a summary of the simulations used including the resolution used to resolve halos in the semi-numerical code, while the last column shows the resolution used in the Lyα calculation.

All simulations use halos with masses down to $10^8 M_\odot$ corresponding to a minimum virial temperature of $T \sim 10^4$ K. Simulation **S1** cannot resolve halos down to these mass scales and we populate the remaining halos in each (empty) cell using a Poisson sampling biased with the underlying density field, giving a mass function consistent with N-body simulations once non-linear corrections are applied, as described in [26].

Radiation sources are prescribed and star formation rates calculated using the halo model described in [32]. We consider only Population II stars from starbursts [28] as contributing to the ionizing photon budget. In order to calculate the Lyα coupling in a given cell, the algorithm assumes that the scattering rate is proportional to the total Lyman series flux arriving in that cell (see again [26]). This Lyman series flux is obtained through a 3-dimensional integration of the comoving photon emissivity, $\epsilon(\mathbf{x}, \nu, z)$ (defined as the number of photons emitted at position $\mathbf{x}$, redshift z and frequency ν per comoving volume, per proper time and frequency), which is assumed proportional to the star formation rate:

$$\epsilon_{\alpha}(\mathbf{x}, \nu, z) = \mathrm{SFRD}(\mathbf{x}, z)\epsilon_b(\nu) , \tag{3}$$

where $\mathrm{SFRD}(\mathbf{x}, z)$ is the star formation rate density from the simulation (in terms of the number of baryons in stars per comoving volume and proper time). As stated above, in the simulation S3 we used the star formation rate obtained from the [33] simulation. $\epsilon_b(\nu)$ is the spectral distribution function of the sources (defined as the number of photons per unit frequency emitted at ν per baryon in stars). We assumed a power law model for $\epsilon_b(\nu)$:

$$\epsilon_b(\nu) = A\nu^{\alpha}, \tag{4}$$

between $\nu_{\alpha}(10.2\,\mathrm{eV})$ and the Lyman limit frequency (13.6 eV). This will allow us to calculate how sensitive the 21 cm power spectrum is to changes in the parameters of the emission model.

We take $\alpha = -0.9$ and A such that the integration between the Lyα and Lyman limit frequencies gives a total emission of 20,000 photons per baryon. These numbers are based on the expected PopII spectra from [28]. Note that the parameter A will be totally degenerate with the star formation rate efficiency in our model. Since the initial mass function (IMF) is likely to be dominated by massive stars, this single power law model is likely to be a good description of the emission spectra (see e.g. [14]), although a broken power law can sometimes provide a slightly better fit [21].

Since the details of the galaxy emissivity depend upon the assumed initial mass function (IMF), which is poorly known [2], these values are subject to considerable uncertainty. Constraints would therefore provide useful information about the properties of the first galaxies. These uncertainties make the mapping between the mean value of the Lyα coupling $\langle x_{\alpha} \rangle$ and redshift uncertain. We will therefore describe the shape of the Lyα power spectrum for a given $\langle x_{\alpha} \rangle$ and caution that the redshift at which that actually occurs may be very different from our fiducial model.

2.2 Identifying the Lyα Epoch

In principle, there are several contributions to the 21 cm brightness temperature and full modeling of the signal would be required to disentangle the part due to Lyα

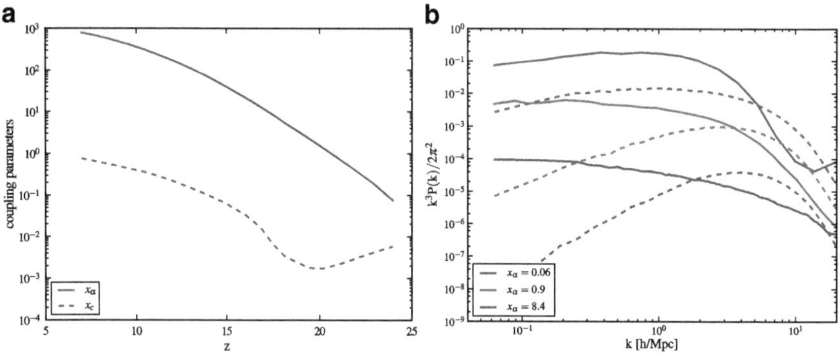

Fig. 2 (**a**) The average x_α and x_c coupling parameters as a function of redshift for the S2 simulation. The Lyα coupling dominates over collisions at all redshifts considered. (**b**) The fluctuations from Lyα, A_α (*solid*) and the gas temperature, $1 - \frac{T_\gamma}{T_K}$ (*dashed*), contributing to the 21 cm brightness temperature. Note that both quantities are normalized

fluctuations and obtain parameter constraints on the first galaxies. However, as we can see from Fig. 2a, b, both the fluctuations in the collisional coupling and in the gas temperature should be subdominant at the redshifts where the Lyα fluctuations are strongest (when $\langle x_\alpha \rangle < 1$ and $k < 1\, h/\mathrm{Mpc}$) making the analysis of the signal more straightforward (the ionization is very small at these redshifts, so that their contribution to the fluctuation power is completely negligible).

The question then is if whether we can actually identify this epoch from observations rather than theoretical guesses. The rms of the signal ($\langle(\delta T_b - \langle\delta T_b\rangle)^2\rangle$) can provide that information and should be easier to measure with interferometers. As we can see in Fig. 3a at $z > 25$ the signal is initially small since coupling to the gas temperature is low. As the Lyα coupling increases, the spin temperature approaches the gas temperature and the rms of the signal increases, not only due to fluctuations on the Lyα field but also because the factor $A_\alpha\left(1 - T_\gamma/T_K\right)$ is increasing (in absolute terms) since the gas is still cooling adiabatically. Once the gas temperature begins to increase due to X-ray heating the signal decreases and we see the turning point at $z \sim 16$ (the transition from absorption to emission occurs later at $z \sim 13$. Finally, the gas is heated far above the CMB temperature and the signal plateaus until reionization finally causes it to die away at $z \sim 7$.

Note that as X-rays start heating the IGM, the signal does not drop immediately since this heating is inhomogeneous and there will be some region still cooling adiabatically as well as temperature fluctuations contributing to the rms. Therefore we see a small plateau at the peak of the signal starting at $z \sim 19$. The conclusion is that it should be safe to neglect the contributions of the X-ray heating to the 21 cm signal between the points when the rms (or the average) starts rising at $z \geq 24$ and before we reach the plateau at the maximum $z \leq 19$ ($x_\alpha \sim 0.9$), at least on large scales where the Lyα contribution dominates over the temperature as we can see in Fig. 2b.

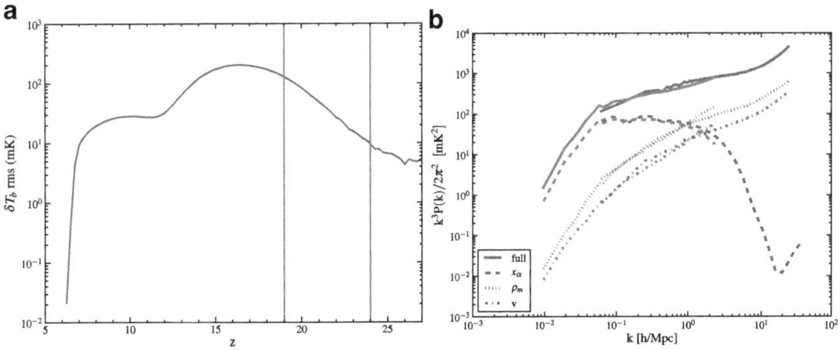

Fig. 3 (**a**) $\sqrt{\langle(\delta T_b - \langle\delta T_b\rangle)^2\rangle}$ (rms) of the signal as a function of redshift for simulation S2. The Lyman alpha coupling saturates around $z \sim 19$. *Red lines* indicate the region where we can safely ignore fluctuations from collisions and X-ray heating. (**b**) The 21 cm power spectrum and its main contributions at $\langle x_\alpha \rangle = 0.4$ ($z \sim 20.25$) using simulation S1 (*red*) and simulation S2 (*blue*). *Solid* – all contributions included; *dashed* – Lyα only; *dotted* -matter density; *dot-dashed* – velocity fluctuations only

In this case, the equation simplifies and we can write the full 21 cm signal as:

$$\delta T_b(\nu) = C(z)(1+\delta)(1+\delta_{A_\alpha})(1+\delta_{A_\nu})(1+\beta\delta), \quad (5)$$

where

$$C(z) \approx 23\langle 1 - \frac{T_\gamma}{T_K}\rangle \langle A_\alpha \rangle \langle A_\nu \rangle \left(\frac{0.7}{h}\right)\left(\frac{\Omega_b h^2}{0.02}\right)\left[\left(\frac{0.15}{\Omega_m h^2}\right)\left(\frac{1+z}{10}\right)\right]^{1/2} \text{ mK} \quad (6)$$

with $A_\nu = \left(\frac{1}{1+1/H\, dv_r/dr}\right)$ ($\langle A_\nu \rangle \sim 1$) and $\beta = \frac{2}{3}\frac{\bar{T}_K}{\bar{T}_K - T_\gamma}$. Figure 3b shows the contribution to the full 21 cm signal from each of the terms considered on (5) using simulations S1 and S2. Again we see that the Lyα term dominates over large scales $k < 1$ h/Mpc, which makes it all important to be able to generate large volume simulations (>100 Mpc/h) and give us confidence that it should be possible to extract the Lyα signal over the redshift range proposed without confusion from other contributions.

3 Measurements with the SKA

3.1 Noise Power Spectrum

Before we go into the details of the experimental setup, we quickly review the noise calculation that goes into our analysis. Following [5, 17, 18, 27], the expected error $\Delta P(k,\theta)$ on the measurement of the 3-d power spectrum of the signal $P_S(k,\theta)$

with noise $P_N(k, \theta)$ is given by

$$\Delta P(k, \theta) = \frac{1}{\sqrt{N_m(k, \theta)}} \left[P_S(k, \theta) + P_N(k, \theta) \right] , \tag{7}$$

where we are assuming that the power spectrum depends only on the moduli (k) of the vector **k** and on the angle θ between **k** and the line of sight. $N_m(k, \theta)$ is the total number of modes in **k** space contributing to the measurement (note that the sum is only done over half the sphere). The noise power spectrum is given by

$$P_N(k, \theta) = r^2 y \frac{\pi \lambda^2 D_{max}^2 T_{sys}^2}{t_0 A_{tot}^2} , \tag{8}$$

where A_{tot} is the total collecting area of the telescope, D_{max} is the maximum baseline assuming that the baseline density distribution is constant on the $u - v$ plane (Fourier dual of the angular coordinates on the sky), $r(z) = \int_0^z cH^{-1} dz'$ is the comoving distance to redshift z, $y \equiv \lambda_{21}(1 + z)^2/H(z)$ is a conversion factor between frequency intervals and comoving distances ($H(z)$ is the Hubble expansion rate and $\lambda_{21} \approx 21.1$cm the rest frame wavelength of the 21 cm line) and the system temperature was modeled by

$$T_{sys}(z) = 50 + 60 \left(\frac{1 + z}{4.73} \right)^{2.55} \text{K} \tag{9}$$

and is dominated by the sky (galactic synchrotron) temperature at the redshifts of interest.

3.2 Experimental Setup

When its completed at around 2,020 the SKA will have an observational window between 70 MHz and 10 GHz and it will be the first radio-interferometer with enough collecting area necessary to probe the high redshift universe ($z > 15$).

The SKA will probably be made up of three different instruments [7, 29]: SKA-low, tailored for the low frequency signal between 70 MHz to 450 MHz, probably made up of sparse Aperture Arrays so that the collecting area scales as λ^2; SKA-mid between 400 MHz and 1.4 GHz, using dense Aperture Arrays; and SKA-high, using dishes for frequencies above 1.2 GHz (note the small overlap between ranges). In this paper, we will assume a low frequency "SKA type" experiment with a setup capable of probing the high redshift, pre-reionization epoch.

The analysis in the previous section showed that in order to measure the imprint of the first galaxies on the 21 cm signal, we need to observe at least up to $z = 20$, corresponding to a frequency of $\nu \sim 68$ MHz. Allowing for some flexibility, we consider an experiment that would be able to measure frequencies down to 60 MHz.

Taking into account the design reference for SKA, we assumed a sensitivity of 4,000 m^2/K at 70 MHz (which in fact requires a total collecting area of $\sim$14 km^2, well above the 1 km^2 used to baptize the telescope) and scaling as λ^2 around these frequencies. In order to allow for designs with lower sensitivities, we also consider instruments with 20% and 10% of the total collecting area of the design reference SKA. For the distribution of antennas, the SKADS[5] reference design assumes they would be collected in 250 stations of 180 m diameter each, with 66% in a 5 km diameter core and the rest along 5 spiral arms out to a 180 km radius.

The signal we are looking for dominates on scales $k \leq 1$ h/Mpc, requiring baselines up to $\sim$10 km for proper sampling at the relevant frequencies, e.g. a resolution of 0.76 arcmin at $z = 20$ ($k_{max\perp} \sim 1.8$ h/Mpc – higher resolution can be achieved along the frequency direction although the error would be much larger since we would only have modes along the line of sight). Taking a reasonable setup, we assumed that 70% of the total collecting area would be concentrated within this core of 10 km in diameter. We do not use the rest of the collecting area in the noise calculation, assuming instead this would be used for point source removal and calibration. The same applies when the total collecting area is only 10% and 20% of the reference design.

For the field of view (FoV) we used 10×10 deg^2, which, taking $z = 20$ as reference, sets a resolution of roughly $dk_\perp \sim 4.5 \times 10^{-3}$ h/Mpc. Assuming that, in the core, correlations within stations are also possible, we considered a minimum baseline of 50 m giving $k_{min\perp} \sim 9 \times 10^{-3}$ h/Mpc which is enough to probe the scales of interest. We assumed that each measurement was done with an interval of 4 MHz and a resolution of 0.01 MHz, giving $dk_\parallel = 0.08$ h/Mpc and $k_{max\parallel} = 16.6$ h/Mpc. Note that the total instantaneous bandwidth should be much larger than this ($\sim$50 MHz), so we can measure several redshifts at the same time. Finally, the total integration time was taken to be 1,000 h (see Table 2 for a summary of the experimental setup).

We used the error given by (7) for the power spectrum assuming a distribution for the antenna/stations of the telescope such that a constant baseline density is obtained. Although this is difficult to achieve in practice since the density of stations normally decreases from the center, it turns out that this configuration gives the best signal to noise in terms of the brightness temperature maps and it should therefore give a good indication of the telescope capabilities to probe the 21 cm signal power spectrum. A final note on foregrounds: they can be the most damaging factor at these low frequencies (together with the calibration issues raised by the ionosphere) although it is expected that if the foregrounds are smooth enough along the frequency direction, they can be simply removed by fitting out a smooth function [20,25]. In this analysis we assume that foreground cleaning was applied on a region of 32 MHz (so as to avoid edge effects on our 4 MHz interval) and was successful on scales smaller than this 32 MHz, e.g. we ignore scales with $k < 0.01$ h/Mpc (see [10]).

[5]http://www.skads-eu.org.

Table 2 Assumed experimental setup[a]

Parameters	Values
Min. freq.	60 MHz
Sensitivity	4,000 m^2/K
Max. baseline	10 km
Min. baseline	50 m
FoV	100 deg^2
Integration time	1,000 h
Freq. interval	4 MHz
Freq. resolution	0.01 MHz

[a]We also considered 20% and 10% of this sensitivity in our analysis. Note that only 70% of the assumed collecting area is used in the core of 10 km diameter

3.3 *Power Spectrum Constraints*

Figure 4a shows the expected error on the three dimensional power spectrum, $P(k)$ (integrated over θ), assuming the above setup. Even if foreground removal affects larger k-modes than assumed here we still have a huge range of measurements available up to $k \sim 7$ h/Mpc and in particular, measurements are quite good on the interval where the Lyα contribution is more important. On very large scales it should be possible to measure the power spectrum even with only 10% of the collecting area. Sample variance dominates on large scales while noise (dashed green line) is dominating on small scales (the number of modes on small scales is limited due to the size of the maximum baselines). Note that we could redistribute the collecting area so that the noise power spectrum followed more closely the signal. The error on large scales is kept small because there are still a large number of measurements N_m due to the high resolution on the $u - v$ space from the large field of view (see Fig. 4a).

With the above measurements of the 3d power spectra, we should be able to constrain some of the characteristics of the first galaxies such as their emission spectra. Although, as we have seen, the Lyα signal dominates on larger scales over other contributions, these constraints will be particularly strong if we fix the cosmology.

Up to now we have been assuming a type II emission spectrum with a total of 20,000 photons per baryon emitted between the Lyα and Lyman limit frequency. If instead we assume that the stars responsible for the Lyα coupling are Pop III, we should have around 5,000 photons emitted per baryon used in stars with a spectral index of $\alpha = 0.29$. Figure 4b shows the results of changing the emission model with the dashed curves corresponding to the PopIII case (note that due to the lower number of photons, these lines also correspond to a lower $\langle x_\alpha \rangle$ of 0.1. The dotted curve corresponds to changing the spectral index to $\alpha = 0.29$ while maintaining the total emission to $5,000$ photons, which will be difficult to distinguish under the

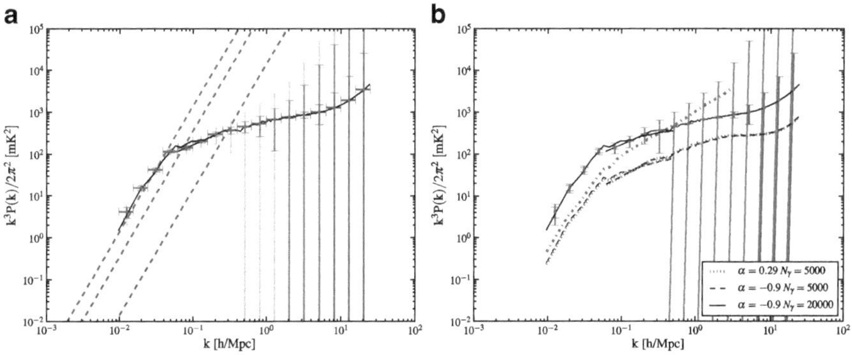

Fig. 4 (**a**) Error on the 3d 21 cm power spectrum at $z = 20.25$ ($(\langle x_\alpha \rangle = 0.4)$) assuming an SKA type experiment. *Solid black* – signal for simulations S1 + S2; *dashed green* – noise power spectrum for the full SKA collecting area, 20 and 10% (increasing amplitude). The expected error taking into account the available number of modes is shown with error bars in *blue*, *red* and *yellow* respectively. (**b**) 21 cm temperature power spectrum for a few Lyman alpha emission models and simulations S1 and S2. *Top black solid lines* correspond to $\langle x_\alpha \rangle = 0.4$ while *bottom dashed/dotted curves* to $\langle x_\alpha \rangle = 0.1$ (all at $z \sim 20.25$). The *green dot-dashed line* uses halos down to $10^6 M_\odot$ and a Pop III type emission spectra. The *error bars* in *blue* corresponds to the full SKA while the ones in *red* to 10% of the collecting area

current noise expectations (the total number of emitted photons per baryon is the most relevant parameter for the Lyα flux). Another possibility is to assume that Pop III stars are also being formed in halos with mass less than $10^8 M_\odot$. The green dot-dashed line in Fig. 4b shows the effect of considering halos down to $10^6 M_\odot$ from simulation S1 at the same redshift. Note that this has a much higher coupling and if we consider higher redshifts so that $\langle x_\alpha \rangle \sim 0.1$ the line will be closer to the dashed one. These calculations show that it should be possible for SKA to measure $\langle x_\alpha \rangle$ from observations of the 21 cm power spectrum at $z \approx 20$. Given a detailed model and small error bars this could be measured directly from the shape of the power spectrum. More generally, if measured over only a small range of wave-numbers, the redshift dependence of $\langle x_\alpha \rangle$ could be extracted by looking at the redshift evolution of the amplitude and slope in a manner analogous to that suggested by [15] for reionization. Note that, although during Lyα domination different models have similar power spectra for the same $\langle x_\alpha \rangle$, its evolution with redshift could be used to distinguish these models.

Measurements of $\langle x_\alpha \rangle$ in a number of different redshift bins would be instrumental in constructing a "21 cm Madau plot" [16]. The Madau plot shows the evolution of the star formation rate as a function of redshift and producing such a plot in the very early Universe would be a major accomplishment of SKA. Measurements of $\langle x_\alpha \rangle$ do not give this directly, since in our model $\langle x_\alpha \rangle$ is determined by the product of the star-formation rate and the emissivity of the galaxies. Breaking this degeneracy observationally will be difficult and would lead to a plot whose absolute normalisation was uncertain, but whose shape tracked the star-formation rate (assuming no evolution of the galaxies). However, for many models of early galaxy formation [6, 14] the variation in the number of Lyα photons produced per baryon is only a factor of two or so. Thus with a reasonable theoretical prior a useful measure of the star-formation history could be established.

4 Redshift Space Distortions

The signal we are trying to probe is actually anisotropic due to the redshift space distortions set by the peculiar velocity of the HI gas, which is already being taken into account in (5). Using a linear approximation, $A_\alpha \approx \frac{\langle x_\alpha \rangle}{1+\langle x_\alpha \rangle}\left(1+\frac{1}{1+\langle x_\alpha \rangle}\delta_{x_\alpha}\right)$ and $A_\nu \approx 1 - \frac{1}{H}\frac{dv_r}{dr}$ so to first order the fluctuation in T_b becomes

$$\delta T_b(\nu) \approx C(z)\left(1 + (1+\beta)\delta + \frac{1}{1+\langle x_\alpha \rangle}\delta_{x_\alpha} - \frac{1}{H}\frac{dv_r}{dr}\right), \tag{10}$$

The peculiar velocity field originates from inhomogeneities in the density field, so that in Fourier space we can use Kaiser's approximation $\frac{1}{H}\frac{dv_r}{dr}(\mathbf{k}) = -\mu^2\delta(\mathbf{k})$ [12] to write the velocity as a function of its dependence in $\mu = \cos(\theta)$. This linear approximation in the velocity contribution allows the brightness temperature power spectra to be separated in only three powers of μ:

$$P_{T_b}(k,\mu) = \mu^4 P_{\mu^4}(k) + \mu^2 P_{\mu^2}(k) + P_{\mu^0}(k), \tag{11}$$

with

$$P_{\mu^4}(k) = C^2(z)P_\delta(k), \tag{12}$$

$$P_{\mu^2}(k) = C^2(z)\left[2(1+\beta)P_\delta(k) + \frac{2}{1+\langle x_\alpha \rangle}P_{\delta\delta_{x_\alpha}}(k)\right], \tag{13}$$

$$P_{\mu^0}(k) = C^2(z)\left[(1+\beta)^2 P_\delta(k) + \frac{1}{(1+\langle x_\alpha \rangle)^2}P_{\delta_{x_\alpha}\delta_{x_\alpha}}(k) + \frac{2(1+\beta)}{1+\langle x_\alpha \rangle}P_{\delta\delta_{x_\alpha}}(k)\right], \tag{14}$$

where the power spectra $P_{ab}(k)$ for generic quantities a and b is defined as:

$$(2\pi)^3\delta_D^3(\mathbf{k}-\mathbf{k}')P_{ab}(k) \equiv \frac{\langle a(\mathbf{k})b(\mathbf{k}')^*\rangle + \langle b(\mathbf{k})a(\mathbf{k}')^*\rangle}{2}. \tag{15}$$

This decomposition can be particularly useful at the high redshifts we are considering, where, as can be seen from Fig. 3b, the contribution from the velocity compression can be relatively strong. By fitting the observed signal to the above polynomial (11) at each k, we can in principle separate the cosmological and astrophysical contributions to the brightness temperature fluctuations [3, 4]. One can use P_{μ^2} and P_{μ^0} to learn about the first sources of radiation while the P_{μ^4} term could be used to measure the matter power spectrum directly, with no interference from any other sources of fluctuations.

4.1 Constraints on $P(k, \mu)$

The redshift space distortions introduce an anisotropy in the 21 cm power spectrum that, in the linear regime, depends on k and even powers of μ. More generally non-linearities in the velocity field will introduce terms with more complicated dependence [31].

We can then obtain P_{μ^0}, P_{μ^2} and P_{μ^4} by fitting equation (11) to the measured $P(k, \mu)$ for $k \leq 1.0$ h/Mpc. This in turn should allow us to obtain direct constraints on a combination of $P_{\delta_{x\alpha}\delta_{x\alpha}}$ and $P_{\delta\delta_{x\alpha}}$. For small k there are fewer bins so the uncertainty in P_{μ^n} is high even if for those scales where the errors in $P(k, \mu)$ are small. Figure 5a shows the expected errors calculated using a Fisher matrix approach. We see that it should be possible to measure with reasonable accuracy both the P_{μ^0} and P_{μ^2} terms which can already give us interesting constraints on the Lyα field. In fact, if we fix the cosmology (e.g. P_δ), then we should be able to extract information from $P_{\delta\delta_{x\alpha}}$ and $P_{\delta_{x\alpha}\delta_{x\alpha}}$ separately. On the other hand, the P_{μ^4} will be hard to measure, which will make the extraction of cosmological information more difficult, simply because it is an order of magnitude smaller than the other terms on the scales we are focusing here.

In order to improve this last constraint we would have to increase the collecting area of the instrument and the frequency interval used (4 MHz in this case) so to have more modes along the line of sight (note however that this will lead to cosmological evolution of the signal along the frequency bin, as already discussed). The SKA should be able to measure modes up to $k \sim 10$ h/Mpc which could have higher values of P_{μ^4}, however for $k > 1$ h/Mpc the angular decomposition is no longer valid and the separate measurements of the $P(k, \mu)$ terms will be of little use.

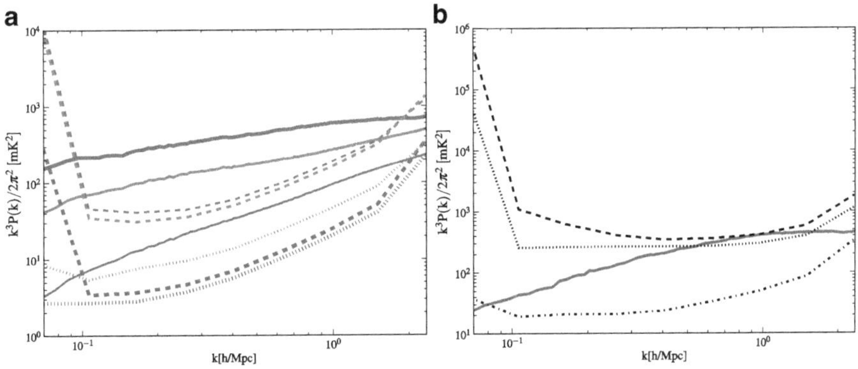

Fig. 5 (**a**) Contributions for the 21 cm temperature power spectrum using simulations (*solid lines*), built using simulation S1 for $\langle x_\alpha \rangle = 0.4$ ($z \sim 20.25$). *Dashed lines*: expected errors. *Dotted lines*: expected errors assuming that P_{μ^4} is known. *Solid lines* from *top* to *bottom* are P_{μ^0} (*green*), P_{μ^2} (*red*) and P_{μ^4} (*blue*). (**b**) Power spectra of $C^2(z)\,(1/(1+\langle x_\alpha \rangle))^2\, P_p(k)$ (*solid line*), built using simulation S1 for $\langle x_\alpha \rangle = 0.4$ ($z \sim 20.25$). Expected error (*dashed line*) and expected error assuming that P_δ is known (*dotted line*), so the error in P_{μ^4} comes only from $C(z)$. *Dot-dashed line* assumes that $C(z)$ is known to high accuracy

In this case, it might be better to just try to fit the parameters of the model directly to the averaged power spectrum $P(k)$ using the simulations (as shown in Fig. 4b). If on the other hand we assume we already know P_{μ^4} with reasonable accuracy then the errors on P_{μ^2} will improve considerably (Fig. 5a) and any degeneracies between P_{μ^0} and P_{μ^2} will be broken.

Finally, by measuring the $P(k, \mu)$ terms, we can in principle use an estimator to probe the properties of the first sources of radiation directly through the Lyα field, without having to go through a full, model dependent, parameter fit to the data. This is based on the decomposition proposed in [3], valid in the linear regime. In that case, δ_{x_α} can be expressed as a function of its dependence on effects correlated and uncorrelated with δ as:

$$\delta_{x_\alpha}(\mathbf{k}) = W(k)\delta(\mathbf{k}) + \delta_p(\mathbf{k}). \tag{16}$$

Note that this decomposition is actually quite general and the main assumption here is the linearity on δ. δ_P is the Poisson contribution that arises from the statistical fluctuations in the number density of the rare first galaxies (galaxies being discrete objects and therefore not tracking the continuous density field perfectly) and which to first order is uncorrelated with the density fluctuations. The window function $W(k)$ expresses the effects by which δ_{x_α} is related to density fluctuations [4]. Using the parametrization of δ_{x_α} we can write $P_{x_\alpha}(k) = W^2(k)P_\delta(k) + P_p(k)$.

According to our simulations, the Poisson contribution dominates for $k >$ 0.2 h/Mpc so that there is just a small interval where we can expect to probe this term before non-linear effects correlated to the density field become relevant.

For large enough scales, we can use the observational data to obtain the Poisson contribution uncorrelated with δ:

$$P_{un-\delta}(k) = P_{\mu^0}(k) - \frac{P^2_{\mu^2}(k)}{4P_{\mu^4}(k)} = C^2(z)\left(\frac{1}{1+\langle x_\alpha\rangle}\right)^2 P_p(k). \tag{17}$$

Figure 5b shows the power spectra of the Poisson contribution (17) and the expected errors associated with this measurement. As expected, due to the high error in P_{μ^4}, the measurement of this Poisson term will be quite difficult even with the assumed experimental setup (dashed line). If we assume that the matter power spectrum is known with reasonable accuracy from CMB data and galaxy surveys, then we can combine the measurement of the P_{μ^4} term at several k to constrain $C^2(z)$ (12), e.g. the mean brightness temperature on the sky. Unfortunately the constraint is still large, $C^2(z) \approx (6.42 \pm 5.60) \times 10^3 \text{mK}^2$, which translates into large errors for the Poisson term (dotted lines). Nevertheless, a detection will still be possible, with a signal to noise of 2.6 (dashed line) and 3.8 (dotted line). If we further assume that we combine this result with the measurements from a global experiment in order to measure the mean brightness temperature with high accuracy, then we can extract all the relevant power spectra reasonably well (dot-dashed line).

5 Conclusions

In this paper, we have made use of a recently developed fast semi-numerical code to explore the possibility for 21 cm observations of the time of the first galaxies with SKA. We demonstrated that this code allows the simulation of the full range of scales likely to be accessible to observations. Here we have focused on the impact of 21 cm fluctuations from large fluctuations in the Lyα coupling, which arise due to the clustering of the rare first galaxies and considered different emission models.

Measurement of these fluctuations would provide insight into the formation of the very first galaxies and would be highly complementary with the next generation of large optical/IR telescopes. We have shown that SKA-pathfinders with ~10% of the full collecting area should be capable of making a statistical detection of the 21 cm power spectrum at redshifts $z \leq 20$. With the full SKA sensitivity this detection would become a measurement allowing astrophysical properties of the first galaxies to be determined.

Observations with an SKA-like instrument would enable the determination of $\langle x_\alpha \rangle$ as a function of redshift from the amplitude and shape of the 21 cm fluctuations. Even though there is a strong degeneracy between the star formation rate and the UV spectral properties of the galaxies themselves, these observations would enable constraints to be placed on the star formation rate in the earliest generations of galaxies. Although crude, the resulting "Madau plot" would be very useful for understanding the formation processes of the first galaxies.

Moving beyond the angle-averaged power spectrum, we have investigated the use of redshift-space distortions to separate out different components of the power spectrum as suggested by [4]. These measurements are difficult due to instrumental limitations and are further complicated by non-linearities on small scales, but with the SKA observations will be able to use this separation to achieve important astrophysical constraints.

Redshift space distortions further offer the possibility of extracting components of the power spectrum that do not correlate with the density field. We have shown that a detection is possible and if strong assumptions about the underlying cosmology are possible and combined with information about the mean 21 cm signal, then these fluctuations may be picked out clearly by SKA.

These results illustrate the potential of 21 cm observations to shed new light on the astrophysics during the pre-Reionization epoch. In the future, our improved knowledge of cosmological parameters will provide a firm foundation to pick out the details of galaxy formation in the early Universe. SKA and pathfinders capable of observing at frequencies $\nu \leq 100$ MHz will begin to access this interesting period and transform our understanding of the cosmic dawn.

Acknowledgements This work was partially supported by FCT-Portugal under grants PTDC/FIS/66825/2006 and PTDC/FIS/100170/2008.

References

1. Baek, S., Di Matteo, P., Semelin, B., Combes, F., Revaz, Y.: Astron. Astrophys. **495**, 389–405 (2009)
2. Barkana, R., Loeb, A.: Phys. Rep. **349**, 125 (2001)
3. Barkana, R., Loeb, A.: Astrophys. J. **624**, L65 (2005)
4. Barkana, R., Loeb, A.: Astrophys. J. **626**, 1 (2005)
5. Bowman, J.D., Morales, M.F., Hewitt, J.N.: Astrophys. J. **638**, 20 (2006)
6. Bromm, V., Larson, R.B.: Annu. Rev. Astron. Astrophys. **42**, 79 (2004)
7. Faulkner, A.J., et. al.: SKADS Memos (2010), http://www.skads-eu.org
8. Field, G.B.: Astrophys. J. **129**, 536 (1959)
9. Furlanetto, S.R., Oh, S.P., Briggs, F.H.: Phys. Rep. **433**, 181 (2006)
10. Harker, G., Zaroubi, S., Bernardi, G., Brentjens, M.A., et al.: Mon. Not. R. Astron. Soc. **405**, 2492 (2010)
11. Hirata, C.M.: Mon. Not. R. Astron. Soc. **367**, 259 (2006)
12. Kaiser, N.: Mon. Not. R. Astron. Soc. **227**, 1 (1987)
13. Komatsu, E., Smith, K.M., Dunkley, J., et al.: ArXiv e-prints, 1001.4538 (2010)
14. Leitherer, C., Schaerer, D., Goldader, J.D., et al.: Astrophys. J. Suppl. Ser. **123**, 3 (1999)
15. Lidz, A., Zahn, O., McQuinn, M., Zaldarriaga, M., Hernquist, L.: Astrophys. J. **680**, 962 (2008)
16. Madau, P., Ferguson, H.C., Dickinson, M.E., et al.: Mon. Not. R. Astron. Soc. **283**, 1388 (1996)
17. Mao, Y., Tegmark, M., McQuinn, M., Zaldarriaga, M., Zahn, O., Phys. Rev. D **78**, 023539 (2008)
18. McQuinn, M., Zahn, O., Zaldarriaga, M., Hernquist, L., Furlanetto, S.R.: Astrophys. J. **653**, 815 (2006)
19. Mesinger, A., Furlanetto, S., Cen, R., ArXiv e-prints, 1003.3878 (2010)
20. Morales, M.F., Bowman, J.D., Hewitt, J.N.: Astrophys. J. **648**, 767 (2006)
21. Pritchard, J.R., Furlanetto, S.R.: Mon. Not. R. Astron. Soc. **367**, 1057 (2006)
22. Pritchard, J.R., Furlanetto, S.R.: Mon. Not. R. Astron. Soc. **376**, 1680 (2007)
23. Pritchard, J.R., Loeb, A.: Phys. Rev. **78**, 103511 (2008)
24. Santos, M.G., Amblard, A., Pritchard, J., et al.: Astrophys. J. **689**, 1 (2008)
25. Santos, M.G., Cooray, A., Knox, L.: Astrophys. J. **625**, 575 (2005)
26. Santos, M.G., Ferramacho, L., Silva, M.B., Amblard, A., Cooray, A.: Mon. Not. R. Astron. Soc. **406**, 2421 (2010)
27. Santos, M.G., Silva, M.B., Pritchard, J.R., Cen, R., Cooray, A.: ArXiv e-prints, 1009.0950 (2010)
28. Schaerer, D.: Astron. Astrophys. **397**, 527–538 (2003)
29. Schilizzi, R.T., Alexander, P., Cordes, J.M., et al.: SKA memo 100, http://www.skatelescope.org
30. Semelin, B., Combes, F., Baek, S.: Astron. Astrophys. **474**, 365 (2007)
31. Shaw, J.R., Lewis, A.: Phys. Rep. **78**, 103512 (2008)
32. Trac, H., Cen, R.: Astrophys. J. **671**, 1 (2007)
33. Trac, H., Cen, R., Loeb, A.: Astrophys. J. Lett. **689**, L81 (2008)
34. Wouthuysen, S.A.: Astron. J. **57**, 31 (1952)

MIX
Papier aus verantwortungsvollen Quellen
Paper from responsible sources
FSC® C105338

If you have any concerns about our products,
you can contact us on
ProductSafety@springernature.com

In case Publisher is established outside the EU,
the EU authorized representative is:
Springer Nature Customer Service Center GmbH
Europaplatz 3, 69115 Heidelberg, Germany

Printed by Libri Plureos GmbH
in Hamburg, Germany